Henry Alleyne Nicholson

Natural History

Its Rise and Progress in Britain as Developed in the Life and Labours of...

Henry Alleyne Nicholson

Natural History

Its Rise and Progress in Britain as Developed in the Life and Labours of...

ISBN/EAN: 9783337025564

Printed in Europe, USA, Canada, Australia, Japan

Cover: Foto ©berggeist007 / pixelio.de

More available books at **www.hansebooks.com**

NATURAL HISTORY

ITS RISE AND PROGRESS IN BRITAIN

AS DEVELOPED IN THE

Life and Labours of Leading Naturalists

BY

H. ALLEYNE NICHOLSON, M.D., D.Sc.

REGIUS PROFESSOR OF NATURAL HISTORY IN THE
UNIVERSITY OF ABERDEEN

W. & R. CHAMBERS
LONDON AND EDINBURGH
1886

Edinburgh:
Printed by W. & R. Chambers.

PREFACE.

IN the present work the Author has endeavoured to give a brief and general outline of the rise and progress of the science of Natural History in Britain. As the great advances in this and other sciences have been for the most part brought about by individual workers, it has been thought desirable, in consistence with the principle of the Series to which this book belongs, to throw this outline into the form of biographical sketches; and as some of the most important steps in the development of the science of zoology have been effected by foreign investigators, it has been necessary to some extent to pass beyond the limits of our own country.

As the last great epoch in zoology is that marked by the appearance of the 'Origin of Species by means of Natural Selection,' the survey here undertaken ends naturally with the great name of Darwin; obvious reasons rendering it undesirable to attempt any estimate of the scientific work of the great naturalists who are still among us.

It is hardly necessary to add that the present work lays no claim to exhaustiveness. Anything of the nature of a detailed history of the rise and progress of Zoological Science would necessarily appeal to experts only. That which has been attempted here, is to give an untechnical, but not unscientific, account of the principal steps which have marked the development of Natural History in our own country. The object of this volume, as of the Series, is to convey through the biographies of the principal workers, an intelligent conception of the progress and leading principles of the science treated of, so that the unprofessional reader may be placed in a position of knowledge to appreciate some of the great questions which at present occupy the scientific world.

CONTENTS

	PAGE
INTRODUCTION	1
ARISTOTELIAN PERIOD	5
RAY AND WILLUGHBY	21
RAY AND WILLUGHBY—*continued*	37
LINNÆUS AND THE LINNEAN CLASSIFICATION	46
THE GREAT MUSEUMS OF BRITAIN—	
Sir Hans Sloane	64
John Hunter	69
BRITISH ZOOLOGISTS	90
BRITISH ZOOLOGISTS—*continued*	107
The Rev. Gilbert White	110
Alexander Wilson	121
CUVIER	136
RETROGRESSION—Swainson and the Circular Classification	168
BRITISH ZOOLOGISTS—*continued*	183
EDWARD FORBES	192
THE DAWN OF THE EVOLUTIONARY PERIOD—Erasmus Darwin	223
THE TRANSMUTATION OF SPECIES—Lamarck	236
THE DOCTRINE OF PROGRESSIVE DEVELOPMENT—The 'Vestiges of Creation'	264
THE THEORY OF NATURAL SELECTION—Charles Darwin	275
THE THEORY OF NATURAL SELECTION—*continued*	293
INDEX	307

LIST OF ILLUSTRATIONS

	PAGE
DARWIN	*Frontispiece*
ARISTOTLE	5
THE CICADA	19
RAY	21
TANTALLON AND BASS ROCK	25
WILLUGHBY	37
LINNÆUS	46
SIR HANS SLOANE	64
JOHN HUNTER	70
ALEXANDER WILSON	122
FISH-HAWK	129
CUVIER	136
LOWER JAW OF WOMBAT—PELVIS OF A KANGAROO	163
THE FOSSIL OPOSSUM OF MONTMARTRE	165
FORBES	192
MAP SHOWING THE DISTRIBUTION OF PLANTS IN BRITAIN	216
FEET OF FOSSIL EQUIDÆ	305

NATURAL HISTORY.

INTRODUCTION.

As Man, upon any theory of his origin, cannot properly be said to have existed as Man, until he had become possessed of that faculty of reason which constitutes his title to the name of *homo sapiens*, it is not altogether extravagant to say that the study of natural history dates its first beginnings from the time of the first appearance of man upon the earth.

Primitive man, whatever may have been the development of his reasoning powers, was assuredly very indifferently provided with the appliances of modern civilisation. So far as concerns their mastery of the forces of external nature, we may, without much risk of controversy, assume that the early races of men were savages. It would therefore have been indeed strange if man, cast, to begin with, amongst a vast series of living creatures, many of which had the power of influencing his material condition for good or for evil, should have shown himself insensible to their presence, or wholly inobservant of their characters, habits, and modes of life. The contrary must always have been the case. It may well be that the rude Palæolithic men who roamed through the trackless forests of Western Europe, clad in undressed skins, and armed only with roughly chipped flints, would gaze wholly unmoved on the thousand beauties of the world around them. Nature has no emotional side, save for those whose souls are freed from the ever-present necessity of procuring food and raiment, shelter from the elements, and protection against wild beasts.

Precisely the same indifference to the softer aspects of nature, and the same insensibility to its beauties, are shown by modern savages, and, for essentially the same reasons, by the poorest members of civilised communities at the present day. We may take it for granted, however, that, just as existing savages are usually accurately acquainted with the larger animals inhabiting their country, so the early flint-men of Post-glacial Europe must have possessed a minute knowledge of the external characters and habits of such animals as the cave-lion, the cave-bear, the mammoth, and the reindeer. Such accurate knowledge of animals, however, even if wholly confined to an acquaintance with their general appearance and mode of life, is, in truth, the basis of scientific natural history.

It is probable, then, that the beginnings of natural history consisted in the knowledge, which the early races of mankind could not fail to acquire, of all those larger animals which, inhabiting the earth or its waters, were either of value for food, or a source of danger from their

INTRODUCTION. 3

size and ferocity. Apart from this, most early mythologies bear testimony to a primeval and widely-spread belief in the mystical or sacred character of various of the more conspicuous animals with which each aboriginal people might happen to be familiar. Not only were particular animals endowed by popular consent with special qualities, good or evil, but specially human attributes were commonly ascribed to them, or they were even regarded as the companions or the representatives of particular deities.

That this association of certain animals with early religious beliefs was, however, of comparatively late growth, is shown conclusively by the fact that, as a general rule, these primitive myths have a distinctly local colouring; the animals regarded as sacred or symbolic by each people being commonly those indigenous to the region inhabited by that people. Thus, the animals regarded with special veneration, or associated with special deities, among the nations of Central and Northern Europe, are such as the bear, the wild boar, and the wolf; while among the peoples of warmer regions similar supernatural qualities are ascribed to the elephant, the lion, the panther, and the peacock. That there is, nevertheless, some common ground for such beliefs is attested by the fact that the same animals are sometimes found to have been credited with some hidden significance among races now widely remote from one another. Thus, to give a single example, the goose—or, it may be, the swan—is mixed up in various ways with the folklore or religious myths of the Hindus, the Romans, the Greeks, and the Northern European races generally. The extraordinarily wide diffusion of early beliefs as to the mystical characters of certain animals is further attested by the known facts as to the system of 'totemism' among primitive races, or by the almost universal traces which are met with of the strange 'cult' known as 'serpent-worship.'

Another and a most important source of zoological knowledge is that arising from the friendly relations which almost all primitive peoples seem to have established with particular kinds of animals. In many cases—indeed in most—such friendly relations seem to have been formed in times long anterior to written history. Philology, moreover, teaches us that among particular groups of nations—as, for example, among all the main stems which have diverged from the great Aryan stock—the names of particular domestic animals are based upon some common root. We thus are furnished with decisive evidence that the animals so designated were known to the Aryans prior to the commencement of their dispersal. Thus, almost all our most valuable domestic animals, such as the ox, the sheep, the pig, the horse, and the dog, are designated in Sanscrit, Latin, Greek, Gothic, and often in German, English, and other allied languages, by names which can be shown to have originated in the same root-form.

ARISTOTLE.

ARISTOTELIAN PERIOD.

WHEN we leave prehistoric ground, and come to the period of written records, we find ample evidence that the ancients were close observers of nature, although natural history, as a science, had as yet no existence. No reader of the Old Testament can fail to admire the beauty, the fitness, and the power of many of the epithets therein applied to animals. An eminent German critic (Gervinus) has remarked, very unjustly, that the ancients had no pleasure in nature; but the writings of Homer, as of all great poets, are by no means without those felicitous phrases descriptive of animals, which show that their author was an acute observer of living beings. It need not be denied that the branches of learning most cultivated by the classical nations were those of rhetoric and logic, grammar, geometry, and metaphysics; and that what we now know as the 'natural sciences' received comparatively little attention, even from the most learned of the Latin and Greek philosophers, while the fine arts were the object of the most successful and the most general pursuit. Nevertheless, Aristotle,* the father of the modern science of zoology, was a Greek, trained in the schools of the Greek philosophy, and as eminent in those purely speculative subjects in which the Greek intellect had always delighted, as he was in the concrete science of natural history.

As there was no 'science,' properly so called, of natural history anterior to the time of Aristotle, and as he may be regarded as the first who gave a systematic form to zoology, it will be well to consider briefly the condition in which this great philosopher left the science which he founded. This is the more needful, as he had no successor, and, so far as the progress of zoology was concerned, might just as well have lived and worked seventeen or eighteen centuries later than the reign of Alexander the Great. However minute may have been

---

* Aristotle was born at Stagira, in the year 384 B.C. He wrote his great work on the 'History of Animals' (Περι τα Ζωα 'Ιστορια) about 340 B.C. As regards the nature and value of the scientific writings of Aristotle, the following may be consulted: 'Aristoteles Thierkunde,' Jürgen Bona Meyer, 1855; 'Life and Analysis of the Scientific Writings of Aristotle,' George Henry Lewes, 1864; 'Leben, Schriften und Schüler des Aristoteles,' Stahr, 1832; 'Geschichte der Zoologie,' Victor Carus, 1872. Short popular memoirs of the life and writings of Aristotle are those of Macgillivray ('Edinburgh Cabinet Library,' vol. xvi. 1834), and of the Rev. Andrew Crichton ('Naturalists' Library,' vol. iii. 1843).

the knowledge which the ancients possessed as to the external characters or habits of the larger animals which they had domesticated, or which lived in a wild state around them, no *system* of natural history had arisen among them prior to the time of Aristotle. Entire branches of zoological science, as now understood, may be said to have been at this time practically non-existent. Moreover, anything like a methodised study of animal life was necessarily attended with extreme difficulties, in the absence of the improved modern means for the investigation and preservation of organic bodies.

In order, however, more clearly to comprehend Aristotle's relations to the science of natural history, what he accomplished, and what he failed to accomplish, where he succeeded in planting a permanent landmark, and where he deviated from the right line of progress, it is needful to have some clear conception as to the precise scope of what is now understood as NATURAL HISTORY.

Now, natural history, as a whole, is only the aggregate history of all known species of animals. The general elements, therefore, which would constitute a perfect science of natural history are the same as those which would give us a perfect history of any given animal. We have, then, to consider what, precisely speaking, are the points upon which we should require to obtain accurate knowledge, if we wished to give a complete history of any single kind or species of animals, such as the dog, for example, or the horse. These points form the bases of the different divisions of a complete natural history, and are as follows :

The first point, then, which we should have to study in connection with any animal is its actual structure—that is to say, its external characters and the form and arrangement of its internal organs. The study of the configuration of animals constitutes that department of natural history which is technically known as MORPHOLOGY, or the science of *Form*—in other words, what is generally understood as ANATOMY.

Secondly, having acquired an acquaintance with the structure of an animal, it is next necessary to study its vital functions, to discover the uses of its internal organs, and the way in which they work, and to investigate the purposes subserved by its various parts when in action. This constitutes the department of natural history known as PHYSIOLOGY, or the science of *Function*.

In the third place, our knowledge of an animal is very incomplete, unless it embraces an acquaintance with those changes which it undergoes in passing from the germ to the adult condition. The study of these constitutes the important branch of zoology known as EMBRYOLOGY, or the science of *Development*.

In the fourth place, it is necessary to know of each species of animals what are its relations to the world we live in. We have to inquire where the species now lives, what are the conditions under which it exists, and so forth; and the study of these points constitutes a special branch of natural history, which is termed the GEOGRAPHICAL DISTRIBUTION OF ANIMALS.

Again, fifthly, we should find that each species of animals has had a *history*, extending backwards into the past; and we should have to study the time of the first introduction of the species upon the earth, the duration of its existence as a species, its geographical range in past time, and various other similar points. The study of these questions constitutes another department of natural history, which is known under the name of PALÆONTOLOGY.

In the sixth place, it is necessary to have some clear ideas as to the relations of each species of animals to other species, and as to the place which the species should occupy in the long series of forms of which the entire animal kingdom is made up. The study of this constitutes what is known as the science of CLASSIFICATION or TAXONOMY; and it is, perhaps, *the* branch of natural history which most nearly fulfils the popular notion of what natural history really is. A 'naturalist' is popularly supposed to be a man who can take any animal and give a name to it, and can place it in some particular drawer or pigeon-hole in the great cabinet of nature. Most of the naturalists of the seventeenth and of the earlier half of the eighteenth centuries concerned themselves principally with questions touching classification; not a few of them, in fact, being little more than 'collectors.' While older workers, not unnaturally, were led to give this branch of the science a position of unmerited prominence, there has of late years been shown a tendency unduly to decry the systematic or taxonomic study of natural history, as being a matter of comparatively small moment. In reality, however, it is not possible to study natural history scientifically, unless we start with the foundation of a systematic classification of some kind. Moreover, if we consider that a strictly 'natural' classification embodies the knowledge of the time as to the lines of descent of animals from primordial types, it would seem difficult to overrate the value and importance of the study of classification.

Finally, we should have to study each species of animal from still another point of view, which, however, is indissolubly connected with the study of its palæontological relations and its systematic position. We should have, namely, to inquire into its mode of origin and its history as a *species*. This constitutes what is known as the science of EVOLUTION.

The above seven heads would embrace the principal sorts of knowledge which the naturalist would find it needful to obtain with regard to any particular species of animal, if he wished to write a complete history of it. Natural history as a whole is, therefore, only the aggregate of our knowledge, under all these seven heads, of all known animals, living and extinct. Perhaps, if our means of investigation were thereto adapted, we should have, further, to study each animal as regards its *Psychology*, or from the side of its mental activity, and also as regards its *Teleology*, or the end for which it exists, and the purposes which it subserves in the general economy of nature. At present, however, we have no adequate means for the study of the psychology of animals; and an inquiry into the teleology of animals is necessarily ultra-scientific. It follows from what has been just remarked, that the departments of knowledge requisite to give us a complete acquaintance with any one animal are also the great departments of the science of zoology. Hence, modern natural history consists of the subordinate subjects of the morphology (or anatomy), physiology, embryology, geographical distribution, palæontology, classification, and evolution of animals.

Speaking generally, the progress of natural history, as a science, has corresponded precisely with the extent to which these separate departments have had their existence recognised, their boundaries defined, and their inter-relations explored. In the early days of natural history these subdivisions had only a very partial existence, or did not exist at all;· and it was for the most part considered enough to acquire a knowledge of the external characters of animals, and of their habits and mode of life, and to give them names by which they could be arranged in some sort of an order. Not only was this the case, but zoology, properly so called, was for long very imperfectly separated from the sciences of botany, geology, and mineralogy, all of which were regarded as forming parts of the general subject of natural history.

This imperfect differentiation of the science of natural history may be said to have prevailed generally up to the middle or nearly the end of the eighteenth century. It is therefore all the more surprising to find that Aristotle, living in the fourth century before our era, and having little or no previously accumulated knowledge to fall back upon, should have been able to form any philosophical conception of the laws of animal life, or to produce a work which should be thought worthy of profound and minute study by the naturalists of the present day.

Aristotle's principal work on natural history, namely his

'History of Animals,' has been the object of extravagant and uncalled-for eulogy; while other critics, as a reaction from this, have shown themselves unduly oblivious of its unquestionable merits. Thus Cuvier,* on the one hand, says of it: 'I cannot read this work without being carried away with wonder. Indeed it is impossible to conceive how a single man was able to collect and compare the multitude of particular facts implied in the numerous general rules and aphorisms contained in this work.' On the other hand, Mr George Henry Lewes,† though admitting that it is 'a stupendous effort' when viewed in comparison with the works which for centuries succeeded it, remarks that 'looked at absolutely, that is to say in relation to the science of which it treats, it is an ill-digested, ill-compiled mass of details, mostly of small value, with an occasional gleam of something better. There is, strictly speaking, no *science* in it at all. There is not even a system which might look like science. There is not one good description. It is not an anatomical treatise; it is not a descriptive zoology; it is not a philosophy of zoology; it is a collection of remarks about animals, their structure, resemblances, differences, and habits. As a collection it is immense. But it is at the best only a collection of details, without a trace of organisation; and the details themselves are rarely valuable, often inaccurate.'

It would be quite out of place to enter here into any detailed analysis of the 'History of Animals,' or of the less famous but not less important treatise 'On the

* 'Histoire des Sciences Naturelles,' 1841, tom. i., p. 146.
† 'Aristotle, a Chapter from the History of Science.' 1864, p. 271.

Parts of Animals.' A few general remarks may be made, however, as to the place which Aristotle holds in relation to the science of natural history.

In the first place, then, Aristotle occupied an attitude towards natural history which in one very important sense agrees with that held by modern scientific workers— namely, in so far as he regarded direct observation as more important than speculation or theory. He cannot be said to have invariably observed this rule strictly; but upon the whole he displays a marvellous love for facts, and also a much greater care in sifting his facts than was usual at the period in which he wrote, or indeed in much later periods. Many of the facts which he records are naturally inaccurately given; others are not facts at all, but mere fictions; while others are so glaringly wide of the truth, and could so readily have been shown to be so,[*] that one can only wonder how they should for an instant have been accepted by a man of his extraordinary sagacity. Nevertheless, Aristotle knew that observation ought to be the guide to all sound scientific investigation, and in this respect he was a notable exception to those philosophers who followed him for many centuries. Moreover, considering the time at which he lived, and the very imperfect methods which he was forced to employ, he must be set down as an observer of pre-eminent powers. This is especially conspicuous in his treatment of certain groups of animals, with which we may suppose him to have had a more intimate personal acquaintance than he could have had with other groups. Thus, his observations upon Cuttle-

---

[*] For example, the statement that the males of the human species, of sheep, of goats, and of pigs, have more teeth than the females.

fishes are not only still valuable, but they embrace certain points which have only been verified and thoroughly elucidated within the present century.

Not only was Aristotle an admirable observer, but he often exhibits a remarkable power of generalisation. Hence he recognised in various instances the existence of biological laws, which have been firmly established only in quite modern times. Thus, he recognised that there often exists a real relationship—what is now called a relationship of 'homology'—between apparently dissimilar parts or organs in different animals. For example, he points out that there exists an agreement of this kind between a claw and a nail, or between the feather of a bird and the scale of a fish. Again, he clearly divined the law of what is now known as the law of the 'correlation of organs'—namely, that certain organs, which are not necessarily in obvious connection with one another, are only found in association with one another, or that certain other organs which, for aught we see, might quite well coexist, are never found in the same animal. Thus, he points out that no quadruped possesses both horns and tusks, and that most animals with horns are two-hoofed; while he adds that a one-hoofed animal with horns (such as the unicorn of heraldry) had never been observed. Similarly he points out that all winged insects which have stings at the fore part of the body (what he calls the sting in these cases consists really of pointed or lancet-like jaws) are two-winged, while those which have the sting at the hinder end of the body are four-winged.

It should be said that Aristotle was essentially what we should now call a 'teleologist'—that is to say, he generally sought not only for the cause but also for the *end* of the natural phenomena which came under his notice. Of course, in many instances the final causes which he assigns for particular phenomena are not only ludicrously inadequate, but are also based upon a totally false view of the facts; as, for example, when he states that 'animals are four-footed because their souls are not powerful enough to carry the weight of their bodies in an erect position,' and that the reason why man (and man only) has flesh on his legs is that the upper part of his body may be rendered lighter, and he may thus be enabled to walk erect. In the present age, the existence of a teleological side to nature is very generally denied; and it may well be that we may never be in a position to investigate the *ends* of any natural structures; so that, *for us*, teleology may have practically no existence. Nevertheless, it is probable enough that this is a point upon which the last word has not yet been spoken.

Apart altogether from the accuracy of his observations, or the value of his theoretical views, the importance of Aristotle's works depends, as has been remarked by Carus,* upon the fact that 'he first created a systematic and scientific method of treating the animal kingdom. This method of treatment not only was fitted to serve, and did serve, as a starting-point for well-grounded investigations directed towards the discovery of new lines of research, or the perfection of old ones; but, above all, it for the first time put zoology in its proper place in the series of the inductive sciences.' The facts which

* 'Geschichte der Zoologie,' p. 71.

he so laboriously accumulated were, naturally, often inaccurate, sometimes wholly baseless. In such branches as anatomy and physiology his knowledge was, of necessity, very imperfect, often entirely erroneous; though in certain fields of physiological inquiry, especially in those relating to development and reproduction, his observations are commonly extremely accurate. Palæontology had, of course, no existence for him; and his acquaintance with the laws of the geographical distribution of animals could not fail to be limited by the limited knowledge which the ancients possessed as to geography itself. Of our modern views as to the evolution of specific forms also, he does not seem to have possessed any foreshadowing. Rather, he held that 'species' had a real existence, and that they were therefore immutable.

Lastly, as a systematist and classifier, Aristotle has undoubtedly been credited with more than his proper due. Upon this point, Mr Lewes's arguments are conclusive. Aristotle has usually been regarded as the first who propounded a general classification of the animal kingdom. The truth is, however, that he propounded no formal or systematic arrangement of animals. As Agassiz* puts it, his work shows 'a total absence of systematic form, of any classification or framework to express the divisions of the animal kingdom into larger or lesser groups. His only divisions are genera and species: classes, orders, and families, as we understand them now, are quite foreign to the Greek conception of the animal kingdom.' It is true that Aristotle divided animals into

* 'Methods of Study in Natural History,' 1864, p. 3.

two great groups, the *Enaima* and *Anaima*, or animals with blood and animals without blood; but he does not use this as the basis of any distinct classification. Under the head of *Enaima*, or animals with blood (that is, with *red* blood), he seems to have included all the animals which we know as the 'Vertebrate Animals,' and he recognises certain distinctions amongst these according as they are viviparous or oviparous, or according to the number of legs which they possessed; but he cannot reasonably be said to have established a *classification* of the *Vertebrata* upon these distinctions. Similarly, the *Anaima*, or animals without blood (that is, with colourless blood, or having, as he thought, a fluid analogous to blood, though not really the same), were understood by Aristotle as comprising the 'Invertebrate Animals;' and he recognised certain groups of these, such as the Shell-fish, the Crustaceans, the Cuttle-fishes, and the Insects; but here also he laid down no regular classification. Upon the whole, therefore, we may accept the verdict of Mr George Henry Lewes* upon this point, that 'zoologists may read a classification in Aristotle's pages, but they do violence to the plain meaning of the text; they disregard context, and piece together from far and wide detached observations never meant to be connected with one another.'

We have dwelt thus at length upon the labours of Aristotle, partly because he is the best, and indeed almost the only, representative of the knowledge possessed by the ancients as to natural history, and partly because it is

* 'Aristotle, a Chapter from the History of Science,' p. 278.

possible to pass at almost a single step from the philosopher of Stagira to John Ray, one of the first of the great British naturalists. With the death of Aristotle, the scientific prosecution of natural history practically came to a close, not for a short time merely, but for a period of many centuries.

The Roman empire, nearly four hundred years later, produced Pliny the Elder, whose name as a naturalist is familiar to every one. In truth, however, Pliny hardly deserves this title at all; since his great work—the 'Historia Naturalis'—is really a kind of cyclopædia, in which little or nothing deserving of the name of 'science' can be detected. It is simply a huge compilation of unassorted facts and fables, principally the latter. From the time of Pliny to the commencement of the sixteenth century, natural history may be said to have been almost at a standstill, and assuredly never even approached the high-water mark left by the researches of Aristotle. With the fall of the Roman empire fell also the learning and the culture of the ancients; and the aptly named period of the 'dark ages' records in its annals few names which would find even a humble place in the zoological temple of fame.

With the dawn of a better time in the beginning of the sixteenth century, natural history did not for long remain unaffected by the general revival of learning. Belon, Rondeletius, Salviani, Conrad Gesner, and Aldrovandus, all well-known naturalists, are names of this period, which bear testimony to a renewed interest in the world of nature. It was not, however, till the early part of the seventeenth century that natural history showed signs of awakening in our own country. The first zoological work ever printed in England is the 'Theatre of Insects' by Moufet,* of date 1634, a translation of which into the vernacular was published twenty-five years later by Edward Topsel. We append an accurate copy of one of the woodcuts of this curious old work, the one selected representing the Cicada.

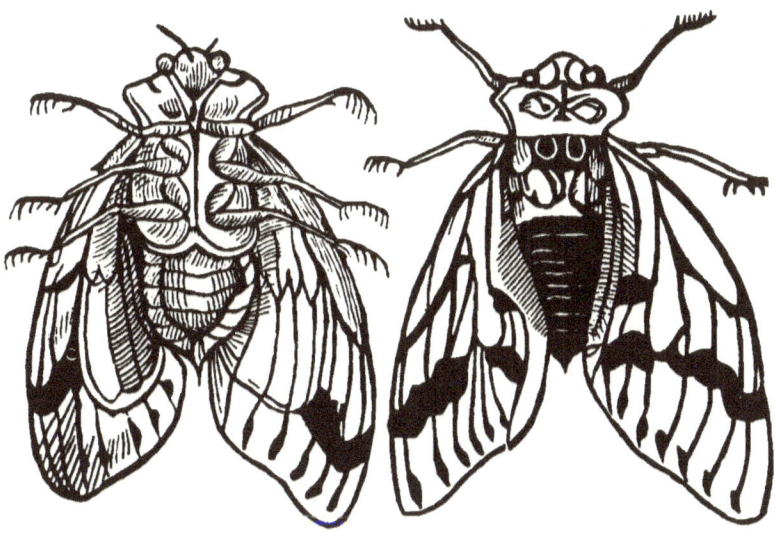

The first work which treated specially of the native animals and plants of Britain is the 'Pinax Rerum Naturalium Britannicarum' of Christopher Merret, published in 1667. This, in addition to animals and plants, embraced fossils also; but it is nothing more than a sort of catalogue or enumeration of the animals and plants known

* 'Insectorum sive Minimorum Animalium Theatrum, olim ab Edoardo Wottono, Conrado Gesnero, Thomaque Pennio inchoatum: tandem Tho. Moufeti Londinâtis, operâ sumptibusque maximis concinnatum, auctum, perfectum t et ad vivum expressis iconibus illustratum. Londini ex officinâ typographicâ Thom. Cotes,' 1634.

to the writer. Indeed, both of these early works are exceedingly rude and imperfect productions, and have no value at the present time. On the other hand, the close of the seventeenth century was rendered memorable in the history of zoology through the labours of several British naturalists, and notably of John Ray and Francis Willughby. Ray may therefore be appropriately selected as the chief representative of the natural sciences in Britain during the pre-Linnean period.

JOHN RAY.

RAY AND WILLUGHBY.

THE seventeenth century has been characterised—of course from the point of view of a naturalist—as 'the dawn of the Golden Age.' The torch of zoological discovery lighted by Aristotle, after flickering fitfully in the hands of his successors for a space, had become extinguished, and the entire domain of natural history had for centuries lain shrouded under the thick darkness of the middle ages. With the commencement of the sixteenth century, the natural sciences participated in the general revival of learning which signalised this period in the history of Europe. Even at the present day, Belon, Rondeletius, and Gesner are something more than merely the shadows of names. In Britain the first pioneers in the renewed exploration of the world of life were little more than mere compilers. The seventeenth century, however, gave origin in Britain to a cluster of eminent men who devoted themselves to the study of zoology and botany, and who for the time being placed England in the first rank as regards the advancement of natural science. The two names which stand out foremost in this cluster are those of the friends and fellow-workers, John Ray and Francis Willughby.

John Ray has been called the 'Aristotle of England,' but he was in reality rather the English Linnæus, his merits as an observer and systematiser being greater than his abilities as a philosopher. Moreover, he was more of a botanist than a zoologist; his purely zoological work being so far blended with that of Willughby, that it is a thankless task to attempt to indicate precisely the share of merit which posterity ought to allot to each of these great men.

John Ray was born on the 29th of November 1628, at the little village of Black Notley, near Braintree, in Essex. He was of humble birth, his father, Roger Ray, being a blacksmith; and he certainly owed comparatively little to his early education, as he was bred a scholar at Braintree School, and has left it on record that he regarded this as a great misfortune. Little is known of his boyish years; but it may be gathered that he was deeply attached to his parents, and especially to his 'most dear and honoured mother.' When he was sixteen years of age he was sent to Cambridge, where he entered at St Catharine's Hall.* In about a year and three-quarters, however, he removed to Trinity College, where he had the advantage of being taught by the well-known Dr Duport, a celebrated scholar of his day.† 'Under this learned tutor,' to use Dr

---

* Ray's name appears in the college books as 'Wray,' he having for a time chosen to write himself in this fashion.

† James Duport was Master of Magdalene College, Dean of Peterborough, and Professor of Greek at Cambridge. He died in 1679. The present writer possesses one of his works entitled 'Threnothriambos,' consisting of the book of Job translated into Greek verse, with a Latin version; together with Proverbs, Ecclesiastes, and Solomon's Song in Greek hexameters. This copy is interesting as bearing upon its

Derham's words, 'Mr Ray so closely applied himself to his studies, that what he missed of at Braintree School, he sufficiently attained to at Trinity College; having acquired great skill in Greek and Latin, and, I have good reason to think, in Hebrew also. Beside which, I find by some of his papers written about that time, that he was very early an excellent orator and naturalist; and upon the account of his great diligence, learning, and virtue, he was soon taken notice of by the College, and at about three years' standing was chosen Minor Fellow of Trinity, on September 8th, 1649, together with his ingenious friend Isaac Barrow; and as Dr Duport had been tutor to both of them, so he used to boast of them, as Mr Ray's fellow-collegian, the late pious and learned Mr Brokesby, informed me, who saith that he, in discourse with Dr Duport, reckoning up several gentlemen of worth that the doctor had been tutor to, the doctor said the chief of all his pupils were Mr Ray and Dr Barrow, to whom he esteemed none of the rest comparable.' *

On taking his degree as Master of Arts, Ray was elected Major Fellow of Trinity, and subsequently filled several highly honourable posts in his college, being successively Greek lecturer, mathematical lecturer, and humanity reader. At this period of his life Ray was not in holy orders; but as the custom of the time was, he nevertheless was in the habit of delivering sermons both in his college and before the university. Such college sermons were usually termed 'commonplaces,'

title-page the signature, 'Tho. Baker, Coll: Jo: Socius ejectus.' It belonged to Thomas Baker, a well-known antiquary of his day, who was one of the 'non-juring' clergy of 1690, and was ejected from his Fellowship in 1716.

* 'Select Remains and Life of Ray, by William Derham, D.D.

a name which doubtless was often fully deserved. In this particular instance, however, we have sufficient evidence that Ray's collegiate 'exercises' were something more than formal platitudes; for some of them (as, for example, his 'Wisdom of God in the Creation') were subsequently published in an enlarged form, and met with the greatest acceptance.

Ray's taste for science seems to have been of early growth; but his first published work was a 'Catalogue of Cambridge Plants,' which was given to the world in 1660. It was merely an alphabetical list of all the plants with which he was acquainted as growing round Cambridge, the localities for each being appended. It was nevertheless much more genuinely scientific than the 'Herbals,' which constituted the chief botanical literature of the day; and it was so far successful, that Ray conceived the idea of preparing a similar catalogue for the whole of England. With this object, he took two journeys (in 1661 and 1662) with his bosom-friend Willughby and some others of his intimates, through portions of England and Scotland, in which he investigated all the objects of scientific or antiquarian interest which came in his way, but more especially the plants and animals. He had previously (in 1658) undertaken by himself a similar journey through parts of England and Wales, and having kept a daily journal, the record of these three journeys was published after his death by Dr Derham. These 'Itineraries,' though not intended by their writer for publication, are highly interesting as giving us a glimpse of the aspect of many well-known places about the middle of the seventeenth century; such observations being indiscriminately mingled with notes upon the plants, animals, or fossils met with, or quaint observations upon men or manners. One might quote endlessly from this *olla podrida* of scientific, antiquarian, and social observations; but it will suffice to take, by way of specimen,

TANTALLON AND BASS ROCK.

Ray's account of his journey from Dunbar to Edinburgh, on which occasion he paid a visit with his companions to the Bass Rock.

'August the 19th, we went to Leith, keeping all along the side of the Fryth. By the way we viewed Tontallon Castle, and passed over to the Basse Island, where we saw on the rocks innumerable of the soland geese. The old ones are all over white, excepting the pinion or hard feathers of the wings, which are black. The upper part of the head and neck, in those that are old, is of a yellowish dun colour; they lay but one egg apiece, which is white and not very large. They are very bold, and sit in great multitudes till one comes close up to them, because they are not wont to be scared or disturbed. The young ones are esteemed a choice dish in Scotland, and sold very dear (1s. 8d. plucked). We eat of them at Dunbar. They are in bigness little inferior to an ordinary goose. The young one is upon the back black, and speckled with little white spots; under the breast and the belly gray. The beak is sharp-pointed, the mouth very wide and large, the tongue very small, the eyes great, the foot hath four toes webbed together. It feeds upon mackerel and herring, and the flesh of the young one smells and tastes strong of these fish. The other birds which nestle in the Basse are these: the scout, which is double-ribbed; the cattiwake, the cormorant, the scart, and a bird called the turtle-dove, whole footed, and the feet red.* There are verses which contain the names of these birds among the vulgar, two whereof are:

> The scout, the scart, and the cattiwake,
> The soland goose sits on the lack,
> Yearly in the spring.

'We saw [some] of the scout's eggs, which are very large and speckled. It is very dangerous to climb the rocks for the young of these fowls, and seldom a year passeth but one or other of the climbers fall down, and lose their lives, as did one not long before our

---

* The 'scout' is the razor-bill; the 'cattiwake' is the kittiwake gull; the 'scart' is the shag; and the 'turtle-dove' is the black guillemot.

being there. The laird of this island makes a great profit yearly of the soland geese taken; as I remember, they told us 130*l.* sterling. There is in the isle a small house, which they call a castle; it is inaccessible and impregnable, but of no great consideration in a war, there being no harbour, nor anything like it. The island will afford grass to keep thirty sheep. They make strangers that come to visit it burgesses of the Basse, by giving them to drink of the water of the well, which springs near the top of the rock, and a flower out of the garden thereby. The island is nought else but a rock, and stands off the land near a mile; at Dunbar you would not guess it above a mile distant, though it be thence at least five. We found growing in the island in great plenty, *Beta marina, Lychnis marina nostras, Malva arborea marina nostras,* and *Cochlearia rotundifolia.* By the way we saw also glasses made of kelp and sand mixed together, and calcined in an oven. The crucibles which contained the melted glass, they told us, were made of tobacco-pipe clay.

'At Leith we saw one of those citadels built by the Protector, one of the best fortifications that ever we beheld, passing fair and sumptuous. There are three forts advanced above the rest, and two platforms. The works round about are faced with freestone towards the ditch, and are almost as high as the highest buildings within, and withal thick and substantial. Below are very pleasant, convenient, and well-built houses for the governor, officers, and soldiers, and for magazines and stores; there is also a good capacious chapel, the piazza, or void space within, as large as Trinity College (in Cambridge) great court. This is one of the four forts. The other three are at St Johnston's, Inverness, and Ayre. The building of each of which (as we were credibly informed) cost above 100,000*l.* sterling; indeed, I do not see how it could cost less. In England it would have cost much more.

'In Edinburgh we went to the principal public buildings. These are: (1) The castle, a very strong building on a precipitous solid rock; it is one of the king's houses, but of no very great receipt; in it are kept the crown and sceptre of Scotland. There was then lying in the castle yard an old great iron gun, which they call Mount's Meg, and some, Meg of Berwick, of a great bore, but the length is not answerable to the bigness. (2) Heriot's Hospital, a square stone building, having a large turret at each corner. It hath very spacious and beautiful gardens, and is well inclosed. There is a cloister on both sides of the court, on each hand as one goeth in, and a well in the middle thereof. At our being there it maintained three-score boys, who wore blue gowns; but they told us it was designed for other purposes. It would make a very handsome college, comparable to the best in our universities. Over the gate, within-side, stands the figure G. Heriot, the founder thereof, and under him this verse,

> Corporis hæc, animi est hoc opus effigies.

(3) The College, for the building of it, [is] but mean, and of no very great capacity, in both comparable to Caius College, in Cambridge. Most of the students here live after the fashion of Leyden, in the town; and

wear no gowns till they be laureat, as 'they call it. At our being there (being the time of the vacancy), there was not a student in town; the premier also, as they call him, was absent in London. In the hall of this College, the king's commissioner, Middleton, was entertained by the citizens of Edinburgh.

'(4) The parliament house, which is but of small content, as far as we could judge, not capable of holding two hundred persons. The Lords and commons sit both in the same room together. There is also a place which they call the inner house, in which sit fifteen lords, chosen out of the house, as it were a grand committee. There is an outer room like the lobby, which they call the waiting-room; and two other rooms above-stairs, where commissioners sit. We saw Argyle and Guthry their heads standing on the gates and toll-booth. At the time we were in Scotland, divers women were burnt for witches, they reported to the number of about one hundred and twenty.'

In 1660, after the Restoration, Ray was ordained; but two years later his direct connection with the Church of England came to an end, in consequence of the passing of the 'Act of Uniformity.' By this act, as is well known, every clergyman was required to declare his assent to everything contained in the Book of Common Prayer, to take the oath of canonical obedience, and to abjure the Solemn League and Covenant; and there is no reason to doubt that Ray, an attached son of the church, would willingly have signed these articles. A declaration was, however, further required, that those who had already signed the Solemn League and Covenant did not lie under any obligation to keep their oath; and to this Ray could not conscientiously subscribe. Rather than do so, he sacrificed his fellowship, and, with it, all his hopes of ecclesiastical preferment; thus setting, along with many other high-minded men, a noble example of adherence to principle, and of disregard for considerations of personal comfort or worldly profit.

Having thus abandoned his career in Cambridge, Ray determined to travel abroad for a time; and he carried this determination into practice early in 1663, in company with his friend Francis Willughby and two of his own pupils. Ray's journeyings on the Continent lasted till March 1666, and took him through a large part of Western and Southern Europe, including Sicily and Malta. The record of his travels, and of the scientific observations which he made thereon, was given to the world in 1673, under the title of 'Observations, topographical, moral, and physiological, made in a Journey through part of the Low Countries, Germany, Italy, and France.' This most quaint and interesting journal abounds in information of all kinds, and upon all sorts of subjects. Observations upon the towns, public buildings, political and social institutions, habits of the people, and natural scenery, are mingled in the oddest way with learned disquisitions upon scientific problems, or descriptions of the animals and plants met with on the journey. Endless quotations might be made, but one characteristic sample will be enough.

Journeying from Nürnberg to Ratisbon, Ray stopped

a day at Altdorf,* which he describes as 'a little walled town, and an university belonging to the Nurenbergers, where there is a pretty physic garden.' He then appends in full, a Latin inscription containing a history of the university, in which 'are maintained thirty-six students at the charges of the city of Nurenberg, which also pays the professors their stipends.' The degrees given by this little university are those of 'doctor of law, physic, and poetry, batchelor of divinity, and master of arts.'

Here Ray was shown some 'serpent-stones, and some petrified cockle and muscle shells;' and he takes occasion thereupon to enter upon a digression of more than a dozen pages as to the nature of 'fossils' in general. In this digression he not only gives a list of the localities known to him as affording fossils, but he discusses all the theories as to their origin, and particularly the two principal rival hypotheses of his day. By one of these theories—which Ray speaks of as the 'general opinion of the antients'—it was held that fossils 'were original the shells or bones of living fishes or other animals bred in the sea;' the ancients in this matter agreeing with modern scientific men. On the other hand, there was the widely spread opinion of those who imagined 'these bodies to have been the effects and products of some plastic power in the earth, and to have been formed after the manner of diamonds and other precious stones, or the crystals of coagulated salts, by shooting into such

---

*Altdorf was at one time a famous university, having come into existence in 1575; but it ceased to exist in 1809, shortly after Nürnberg had been incorporated with the kingdom of Bavaria.

figures.' Ray winds up a long argument upon the merits of these two theories as follows: 'For my own part, I confess, I propend to the first opinion, as being more consonant to the nature of the thing, and could wish that all external arguments and objections against it were rationally and solidly answered.' Having satisfactorily disposed of this knotty point, he then immediately continues the thread of his narrative, as if it had been wholly uninterrupted, by giving a full list of the then professors in the university of Altdorf, together with a statement as to the books which were studied in the different classes.

The two or three years which followed Ray's return home, were occupied by him in all sorts of scientific work, and in travelling through England. Much of his time he spent with his friend Willughby, with whom he carried out at this period a series of well-known experiments upon the ascent and descent of the sap in trees. In 1667, Ray became a Fellow of the Royal Society of London, which had only been incorporated for about five years. Two years later he published his 'Catalogue of English Plants,' which, after going into a second edition, was remodelled, and ultimately appeared in 1690 as the famous 'Synopsis Methodica Stirpium Britannicarum.' So far as botany was concerned, this well-known treatise 'proved the great corner-stone of his reputation.'

Year after year thus passed by, and found Ray still absorbed in his peaceful and uneventful scientific labours, till, in 1672, his friend and coadjutor Willughby was carried off by a fever; 'to the infinite and unspeakable loss and grief,' he writes, 'of myself, his friends, and all good men.' This event materially changed Ray's life for a time at any rate. He had been left by Willughby an annuity of sixty pounds a year, with the charge of superintending the education of his sons, Francis and Thomas, the eldest of these being at this time not four years old. He was also left as Willughby's literary and scientific executor.

Under these circumstances, Ray made Middleton Hall, the seat of the Willughbys, his headquarters; and occupied himself during the next three or four years in the education of the boys intrusted to his care, and in editing and completing Willughby's manuscripts. His work in connection with the latter will be noticed hereafter; but some idea may be formed of his wonderful intellectual activity and power of work, from the fact that, in addition to carrying out the duties above mentioned, he published a 'Nomenclator Classicus' (1672), his work on the Low Countries (1673), and his well-known 'Collection of Unusual or Local English Words' (1673). He also, at the same time, carried on an extensive correspondence with Martin Lister, Mr Oldenburg (the Secretary of the Royal Society), Sir Hans Sloane, Dr Tancred Robinson, and others. At this period, letters were something much more serious and substantial than they are at the present day, and many of Ray's letters to Oldenburg were published in the 'Philosophical Transactions.' They dealt with the most extraordinary variety of subjects, amongst which Dr Derham enumerates, *St Paul's battoons* [peculiar stones found in Malta], 'the Trochites of mushrooms, maize, the bleeding of trees and motion of their sap, spontaneous generation, musk-scented insects, the scolopendra, the acid juice of pismires, the darting of spiders, the anatomy of the porpus, the air-bladder in fishes, the macreuse' [scoter duck], 'and the woodcracker.'* A number of these letters, together with those of his correspondents, were published shortly after Ray's death by Dr Derham; in which the industrious reader will surely discover, as Dr Derham quaintly puts it, 'so entertaining and profitable a variety of curious learning, as will sufficiently compensate for defects, and cause him to think that neither have I cast away my time and pains, nor he his cost.'

Amidst all this press of scientific work, Ray found time for the softer emotions, and in June 1673 he was married, in Middleton Church, to Margaret, daughter of Mr John Oakeley of Launton in Oxfordshire, his wife being some twenty-five years younger than himself. About two years after this event, Ray, who had with his wife been living at Middleton Hall in Warwickshire, removed to Sutton Cofield, a place about four miles off; his tuition of Willughby's sons having now come to a close. Two years later (1677) he again moved his quarters, this time to Falborne Hall in Essex, whence two years afterwards he finally shifted to Black Notley, his native place, in which he spent the remaining years of his life.

From this time onward Ray's life was wholly uneventful, and was occupied in incessant and most fruitful toil at his favourite sciences, and in constant correspondence with his friends. Book after book came from his hands, some botanical, some zoological, some theological. Some

* 'Memorials of Ray,' p. 32.

account of these will be given immediately. In the meanwhile we must hasten to the end. Ray's marriage had not been blessed with children, and the last ten years of his life were embittered by failing health and a painful disorder. To the last, however, he intermitted not in his labours, and was engaged in writing perhaps the best of his zoological treatises (namely his 'History of Insects'), when death overtook him. His last letter, penned on his deathbed, bearing, as Dr Derham says, 'the marks of a dying hand in every letter,' and broken off at the end by reason of failing strength, was to Sir Hans Sloane. It is subjoined below in its entirety. Few who have known what it is to enjoy a long and tried friendship, will read it unmoved.

DEAR SIR—*The best of friends;* these are to take a final leave of you as to this world. I look upon myself as a dying man. God requite your kindness expressed any ways towards me an hundred fold, bless you with a confluence of all good things in this world, and eternal life and happiness hereafter, and grant us an happy meeting in heaven. I am, sir, eternally yours, JOHN RAY.

BLACK NOTLEY, *Jan. 7th*, 1704.

Postscript.—When you happen to write to my singular friend Dr Hotton, I pray tell him I received his most obliging and affectionate letter, for which I return thanks, and acquaint him that I was not able to answer it; or——

On the 17th of January 1704, in his house at Black Notley, Ray died, in the seventy-seventh year of his age.

So passed away a great and good man, who to pre-eminent intellectual gifts united a singularly upright, simple, pure, and lovable disposition. As his biographer Derham has it—'in his dealings no man more strictly just; in his conversation, no man more humble, courteous, and affable; towards God, no man more devout; and towards the poor and distressed, no man more compassionate and charitable, according to his abilities.'

WILLUGHBY.

RAY AND WILLUGHBY
(CONTINUED).

IN estimating Ray's intellectual achievements, and the value of his scientific labours, his published works may be divided into three groups. The first and most considerable of these embraces his botanical treatises. These we must wholly pass over here, though it is upon them that Ray's scientific reputation most largely rests. It is enough to quote in this respect the verdict of Sir James E. Smith ('Rees's Cyclopædia'), that Ray was 'the most accurate in observation—the most philosophical in contemplation—and the most faithful in description, amongst all the botanists of his own, or perhaps any other time.'

The second division of Ray's works includes his theological treatises, of which two attained great celebrity—namely, 'The Wisdom of God in the Creation,' which went through many editions, and his three 'Physico-theological Discourses concerning the Chaos, Deluge, and Dissolution of the World,' of which three successive editions appeared. The first of these is an elaborate and learned survey of nature in general, and of the structure of the body of man and animals in particular, from the teleological point of view. From that point of view, it is a wonderful production for the time in which it was written—clear in conception, full in illustration, elevated in sentiment, and dignified in language. The second of the works just referred to is, in the main, a cosmogony and a theory of geological action.

As was natural at the time he wrote, and with his views, Ray did not wholly eschew hypothetical causes in his endeavour to explain how the earth had reached its present condition. All the geologists of his day relied upon imaginary causes, or invoked hypothetical agencies in their explanations of the formation of the earth. Ray, however, was in this respect more advanced than most of his contemporaries, for he relied upon known physical agencies, whenever it seemed to him possible to account for the presumed course of events by the ordinary and recognised operations of nature. He not only showed a desire to get rid of imaginary causes in his explanation of the creation of the world; but he endeavoured to explain the anticipated dissolution of the world by similar causes. Thus, he drew particular attention to the 'denuding' action of rain, rivers, and the sea, and to the manner in which the dry land is at the present day worn away by these agencies; and he speculates upon the result in the future of the long continuance of this action. The value of Ray's treatise is now, of course, purely historical; but as a piece of philosophical reasoning, it attains a much higher level than its more celebrated contemporary, the 'Essay towards a Natural History of the Earth and Terrestrial Bodies,' by John Woodward (1695).

The third group of the works which the world owes to Ray comprises his zoological treatises. Ray's contributions to the science of zoology are, however, so largely, and in many respects so indissolubly linked with those of his friend Willughby, that it is not possible, even if it were desirable, to consider the two separately. It may therefore be proper to preface this subject by the following brief account of the life of the latter.

Francis Willughby was born at Middleton in Warwickshire, in the year 1635, and was the only son of Sir Francis Willughby. Little is known of his early life, except that he was a most diligent student. In 1653, he entered at Trinity College, Cambridge, graduating as Bachelor of Arts in 1656, and as Master of Arts in 1659. It was at Cambridge that he commenced his lifelong friendship with John Ray, whose pupil he is stated to have been; though on this point there seems to be some doubt. Be this as it may, it is certain that from this time forward Ray and Willughby became the closest of intimates, and that both threw themselves with the utmost ardour into the study of nature. Ray had largely devoted himself to botany, whereas the bent of Willughby's mind was rather to zoology; though the former also extensively investigated animals, and the latter worked at times with plants. Hence, when Ray, being forced, as already narrated, to abandon his life at Cambridge, determined to travel abroad, Willughby agreed to join him; they having formed the design of preparing together a kind of general account of the animal and vegetable kingdom. Ray's own version of this design, and of the part which it was intended that he and Willughby should respectively take in its execution, has been preserved to us by Dr Derham, who writes as follows upon this point:

'These two gentlemen, finding the "History of Nature" very imperfect, had agreed between themselves, before their travels beyond sea, to reduce the several tribes of things to a method; and to give accurate descriptions of the several species, from a strict view of them. And forasmuch as Mr Willughby's genius lay chiefly to animals, therefore he undertook the birds, beasts, fishes, and insects, as Mr Ray did the vegetables. And how each of these two great men discharged his province, the world hath seen in their works; which show that Mr Ray lived to bring his part to great perfection; and that Mr Willughby carried his as far as the utmost application and diligence of a short life could enable him.'

The joint tour, for which the above great design served as an object, lasted, as has been previously seen, from 1663 to the beginning of 1666; but Willughby parted from Ray in the later part of 1664, for the purpose of travelling in Spain. The journal of his Spanish tour was appended to the 'Travels in the Low Countries,' published by Ray in 1673; but it was unfortunate that all the scientific papers written by the two friends while they were together, and describing the animals and plants which they had met with, were lost on their return.

At the close of the year 1665, Willughby lost his father, and succeeded to the family estates. He now took up his abode at Middleton Hall, where Ray was his frequent guest. Willughby now applied himself with the utmost zeal to the execution of his great plan of publishing a systematic work dealing with the whole animal kingdom. Ray was often with him, helping him in ordering the extensive collections which he had accumulated. The two took scientific tours together—making their way at one time as far as Land's End; and they worked together at various scientific researches. Thus for a few years Willughby lived, immersed in his work, and enjoying his quiet home-life. In 1668, he married, but four years later he was attacked with pleurisy and fever, to which he succumbed at the early age of thirty-seven. 'Thus,' as Dr Derham remarks, 'was the world deprived of this great and good man, in his very prime. . . . His example deserves the imitation of every person of estate and honour. For he was a man whom God had blessed with a very plentiful estate, and with excellent parts, capable of making him useful to the world; and accordingly he neglected no opportunity of being so. He did not (as the fashion too much is) depend upon his riches, and spend his time in sloth and sports, idle-company keeping, and luxury; but practising what was laudable and good, and what might be of service to mankind.'

Had it not been for the affectionate care of his friend Ray, the world would not have been in a position to estimate, even in part, what it had lost in Francis Willughby. For, though he had been so long engaged in scientific researches, Willughby, up to the time of his death, had published hardly any of his observations. He had, in fact, published nothing independently, save two or three entomological papers in the Transactions of the Royal Society, of which he was a Fellow. Ray, however, undertook to edit and bring out the mass of scientific notes which Willughby had for the most part 'rhapsodically written in Latin,' and which he had left in preparation for his contemplated work on animals; and this task Ray discharged with the utmost fidelity. Indeed, much controversy, of a wholly profitless nature, has arisen as to the respective share of the original author and of the editor, in the zoological treatises which subsequently appeared under the name of Willughby. We may take it for granted that much of the merit of these treatises was due to Ray, but that their groundwork should be credited to Willughby. On this point, we have Ray's own testimony; who says that, on examining Willughby's manuscripts after his death, he 'found the several animals of every kind, both birds and beasts, and fishes and insects, digested into a method of his own contriving, but few of their descriptions or histories so full or perfect as he intended them.'

The first work of Willughby's which was published under the editorial superintendence of Ray was a systematic treatise on birds, 'wherein all the birds hitherto known, being reduced into a method suitable to their natures, are accurately described.'* This great work was published in one volume, folio, in 1676; it was translated by Ray into English in 1678; and a French edition was published by Salerne in 1767. The descriptive part of this well-known treatise is very good, and it contains excellent accounts of the habits of the birds described. Much cannot be said, however, for the illustrations, so glowingly described in the title. They are mostly poor copies of previously existing figures, and according to Macgillivray—an excellent authority—there are 'not ten figures in the work which bear an accurate resemblance to their originals.' The classification adopted in the work is a purely artificial one, as, indeed, it could not have otherwise been; and it would serve no particular purpose to summarise it or discuss it here. It may be added, however, that in spite of its deficiencies, it is his 'Ornithology' which has mainly conduced to keep alive the memory of Willughby as a naturalist.

Ten years after the appearance of the 'Ornithology,' Ray edited and published a second work of Willughby's, under the title 'Historiæ Piscium, Libri Quatuor,' London, folio, 1686. This famous treatise contained descriptions and figures not only of most of the fishes which had been recorded by previous ichthyologists, but also of numerous

* 'Ornithologia, sive de Avibus, Libri tres: in quibus Aves omnes hactenus cognitæ, in Methodum naturis suis convenientum redactæ, accurate describuntur. Descriptiones iconibus elegantissimis et vivarum avium simillimis æri lucide illustrantur. Totum opus recognovit, digressit, supplevit Johannis Raius.'

types which had come under the observation of the authors in their joint travels on the Continent. Thus Cuvier states that it contained many observations on the fishes of the Mediterranean, which were not to be elsewhere obtained; and one of our highest living authorities (Dr Albert Günther) has stated that 'it is no exaggeration to say that at that time these two Englishmen knew the fishes of the Continent, especially those of Germany, better than any continental zoologist ('Introduction to the Study of Fishes,' p. 8). It is interesting to note, by way of marking the progress of natural history, that while Aristotle is supposed to have been acquainted with about one hundred and fifteen species of fishes, Ray estimates the total number known to him at about five hundred species. At the present day, on the other hand, naturalists are acquainted with about six thousand species of fishes; nearly seven hundred species of fresh-water fishes—a number greater than the whole of the fishes known to Ray—having been recognised as occurring in the single area of tropical South America.

In addition to the 'Ornithology' and 'Ichthyology' of Willughby, Ray published under his own name three *Synopses*, so as to give a brief conspectus of the entire series of the Vertebrate Animals. The first of these appeared in 1693, with the title, 'Synopsis methodica Animalium Quadrupedum, et Serpentini Generis.' This work commences with an introduction dealing with the general characters of animals; but the bulk of the work is a synoptical account of Mammals, Reptiles, and Amphibians. The classification of animals here adopted is of interest, as almost the first attempt at a systematic arrangement of the animal kingdom; and it will be noticed more particularly immediately, in connection with the later Linnean classification. The second 'Synopsis Methodica' appeared in 1713, after Ray's death, and consisted of two parts, one dealing with birds, and the other with fishes, thus completing the series of the Vertebrate Animals.

Lastly, Ray undertook, at the very close of his long and laborious life, to complete a systematic treatise on Insects, which Willughby had intended to write, and had, in fact, sketched out. This work Ray did not live to complete, but it was published after his death, in 1710, at the expense of the Royal Society, under the title, 'Historia Insectorum.' It is unfortunate that this treatise was not accompanied by illustrations, which detracted considerably from the value which it might otherwise have possessed. Under the name of 'Insects,' Ray understood not only the animals now so named, but also the Spiders and Scorpions, the Centipedes and Millepedes, and the true Worms. The classification of the Insects proper which Ray adopted is based essentially upon their 'metamorphoses,' that is to say, upon the changes which the insect undergoes in passing from the egg to the adult condition; and though, as might be supposed, in many respects wholly out of accord with modern knowledge, it nevertheless clearly recognises various natural groups of insects.

LINNÆUS.

LINNÆUS AND THE LINNEAN CLASSIFICATION.

At the time of Ray and Willughby, of which we have just been speaking, naturalists were principally occupied with making collections of animals; with accumulating observations of all sorts in connection with their habits and mode of life; and, above all, with framing classifications. The zoological works of Ray and Willughby were largely concerned with classification; and it may be said that they contain the first systematic classification of the whole animal kingdom which had been attempted up to this time. Their classification of animals was, however, necessarily, altogether of what is called an 'artificial' character; and we may here explain briefly what is meant by an 'artificial' as opposed to a 'natural' classification. Classification is simply the arrangement of a series of objects in some kind of *order*, and its most obvious purpose is to supply a means of identifying and finding any given object in the series. When naturalists first came to investigate the vast series of the animal kingdom, they were at once confronted with the necessity of establishing some arrangement of these, so that they might be able to find out what any new animal was, and to place it in some group. As, however, their knowledge was very imperfect, they naturally adopted obvious and conspicuous characters as the basis of their arrangement, regardless of the fact that such characters are often of little real importance, and may be quite outweighed by other much less readily recognisable features. Ray's classification of animals, which, in a slightly modified form, is here subjoined, is based in this way on a few obvious characters. Ray divided all animals as follows:

I. RED-BLOODED ANIMALS (= *Vertebrata*).
  1. Respiring by lungs, and having a heart furnished with two ventricles—
    A. Viviparous—
      a. Aquatic..........................Cetacea (whales, &c.).
      b. Terrestrial........................Ordinary Mammals.
    B. Oviparous...........................Birds.
  2. Having a heart with a single ventricle—
    A. Air-breathers, with lungs...........Reptiles.
    B. Breathing by means of gills.........Fishes.
II. WHITE-BLOODED ANIMALS (= *Invertebrate animals*).
  1. Of large size .... { Malacia, or Mollusca.
                             Malacostraca, or Crustacea.
                             Ostracoderma, or Testacea.
  2. Of small size.......................Insecta.

As regards the red-blooded or Vertebrate animals, the structure of the ventricle of the heart is taken in the above classification as the primary element of the arrangement. Hence the first thing which Ray himself would have done, in dealing with some Vertebrate animal which he did not know, would have been to examine its heart. If it had a double ventricle, he would have concluded that it must belong to the Cetaceans (whales, dolphins, &c.), the ordinary terrestrial quadrupeds, or the birds. If it had a single ventricle, he would have placed it either among the reptiles (under which name he included both the true Reptiles and the Amphibians), or among the fishes. Supposing the animal to belong to the first series —that is, to the groups with a double ventricle—Ray would have next investigated the method in which it brought forth its young. If it were viviparous, he would have his further choice restricted to the two groups of the Cetaceans and the ordinary Mammals; if it were oviparous, he would be able to place it definitely among the birds. In the case of its being viviparous, the only character which would be consulted would be whether it lived in the water or upon the land. If it were an aquatic animal, Ray would locate it finally among the Cetaceans; whereas it would find a place among the ordinary quadrupeds, if its habits were terrestrial.

Now this classification, though sufficiently convenient when there is no particular object in view other than simply to discover to what great group a given Vertebrate animal is to be referred, is an eminently 'artificial' arrangement. That is to say, the characters used to separate the different groups are to a large extent non-essential ones, and do not express the real relationships of the animals grouped together. Hence, this arrangement separates certain animals which are closely allied to one another, and places others in juxtaposition between which there is no real affinity. Thus, the Cetaceans (whales and dolphins) are separated from the ordinary quadrupeds, with which they agree, in all really essential points, simply upon the trivial ground of their aquatic habits; though Ray places the almost equally aquatic Seals in their right place, and even puts the Manatees (sea-cows) among the ordinary Mammals. Again, the birds, by the structure of their heart, are placed in the same primary division as the Mammals, whereas their true affinities are with the Reptiles, in spite of the incomplete heart possessed by the latter.

The above classification, therefore, does not express the true order of nature, or indicate correctly the real relationships that subsist between different groups of animals. Like all 'artificial' classifications, it is essentially a mere index to the book of nature; and though extremely useful for the purpose of discriminating between different species of animals, it no more expresses the relationships of these species to each other, than a good index would enable one to recognise the connection between the different subjects of a volume.

On the other hand, a classification is a 'natural' one just so far as it *does* express these relationships. On a really 'natural' classification, animals are grouped together according to their true affinities to each other; and as these affinities are manifested in *all* the parts of the organisation of animals, so the characters which are used in such a classification are those of the entire structure,

D

and not merely some single peculiarity. A 'natural' classification, therefore, resembles a subject index to a book, rather than a mere alphabetical index. Of course, no perfect classification would be possible unless our knowledge of all animals were perfect; and as this is not the case, even the best of modern classifications is necessarily, to a larger or smaller extent, an 'artificial' one. Still, naturalists recognise now, that the merit of a classification is in direct proportion to the extent to which it ceases to be a mere arbitrary or convenient grouping, and is based upon the points of structural or morphological agreement between animals, quite irrespective of such secondary matters as the way in which they live.

At the period at which Ray lived, our knowledge of animals was not sufficiently extensive to render the framing of anything but an artificial classification possible. Moreover, there existed at that time no clear and definite system of zoological nomenclature. What was at that time understood as a 'genus' was generally what we should now call a 'family,' or, in many cases, even an 'order' of animals. There was also no fixed method of naming 'species' of animals, so as to clearly indicate by their names what precise place they occupied in their own group. As an example of this, we may take the British species of the Crow family, and contrast their names at the present day with those employed by Ray:

		Ray's nomenclature.	Modern nomenclature.
1.	Raven	*Corvus*	*Corvus corax.*
2.	Carrion-crow	*Cornix*	*Corvus corone.*
3.	Rook	*Cornix frugilega*	*Corvus frugilega.*
4.	Hooded-crow	*Cornix cinerea frugilega*	*Corvus cornix.*
5.	Jackdaw	*Monedula*	*Corvus monedula.*
6.	Chough	*Coracias*	*Fregilus graculus.*

It will be seen from the above that the names which Ray used for the half-dozen British species of 'crows' (in the wide sense of this term) are sometimes single, sometimes double, sometimes treble; and also that the names given to the raven and the jackdaw would not indicate any relationship to the rook, the hooded crow, and the carrion crow. On the other hand, the names used in the modern system for the same birds, are, in the first place, all built upon one system, being binomial, consisting of two names, the second corresponding with a man's Christian name and the first with his surname. We also see that the raven, carrion crow, rook, and hooded crow are closely allied to one another, as they all belong to the single sept or 'genus' *Corvus*.* On the other hand, the chough belongs to a group of crows distinguished from our commoner forms by certain special peculiarities, in which it agrees with two other existing species, and it is therefore removed from the rook and its immediate allies, and placed in the separate genus *Fregilus*.

Ray's classification of animals was, at the time it appeared, the best arrangement of the animal kingdom which had been brought forward. It was not destined, however, to live long; and it was superseded, less than forty years after Ray's death, by the system propounded by the celebrated Swedish naturalist Linnæus, whose life may be briefly sketched here.

Karl von Linné—usually known by the Latinised name of Linnæus—was born at Rashult, in the province of Småland in Sweden, in the year 1707. His father was

---

* Many ornithologists now break up the genus *Corvus* into subordinate groups, which are distinguished by special names.

pastor of Rashult, and Linnæus himself was educated with the intention of entering the ministry; but the strong taste which he early manifested for botanical and other scientific pursuits led to his ultimately entering upon the study of medicine in preference to that of divinity. In pursuance of this end, Linnæus became a student in the university of Lund in 1727, being greatly assisted in his scientific studies by Stobæus, the professor of medicine and botany, in whose house he lived. Being, however, strongly attracted to the university of Upsala by the superior facilities which it offered for the study of the natural sciences, Linnæus abruptly left his friend and patron Stobæus, and betook himself to the more famous seat of learning. Here he found himself reduced to great straits by reason of poverty, and it was only after a year or more passed in extreme indigence, that he was relieved from this condition through the kindness of Professor Celsius, who received him as a guest into his house. He now ardently prosecuted his studies in botany and zoology, and was shortly afterwards appointed assistant to Rudbeck, the professor of botany in the university. It was at this time also that he made the acquaintance of Peter Artedi,* a fellow-student and kindred spirit; and the intimacy thus commenced lasted till the death of the latter in 1735.

In the year 1732, Linnæus was selected by the Royal Academy of Upsala as a properly qualified person to

---

* Artedi devoted himself particularly to the study of fishes, and is best known at the present day by his famous 'Ichthyologia,' which was edited and published by Linnæus in 1738, three years after the death of Artedi. In the close friendship which subsisted between Linnæus and Artedi, and in the services which the former rendered to the latter as scientific executor, there is much to remind us of the relations which existed between Ray and Willughby.

investigate the scientific productions of Lapland. In the discharge of the commission thus intrusted to him, he undertook a toilsome and solitary expedition, which consumed six months' time, and was attended with many hardships, but which was fruitful in scientific results. The next two years of the life of Linnæus may be passed over, as being little more than a record of the struggles of a proud, poor, and irritable genius to wring from his countrymen the recognition and position which his talents deserved, but which were withheld from him on account of his poverty. Being prohibited from lecturing at Upsala, on the ground of his not having taken his academical degree, he ultimately (1734) commenced to give lectures on mineralogy at Fahlun, where he fell in love with his future wife, the daughter of a Dr Moræus. The lady in question had saved about one hundred dollars, which sum she gave to Linnæus, in order that he might take his degree—as was commonly done at that time—at some foreign university. With this end in view, he journeyed to Holland, and graduated as doctor of medicine in the university of Harderwijk. Proceeding to Leyden, Linnæus made the acquaintance of Gronovius,* who induced him to publish (1735) his 'Systema Naturæ.' This work, afterwards so famous, comprised a classification of all natural objects, animals, plants, and minerals; and in this, the first edition, it consisted of only fourteen folio pages.

From Holland, Linnæus journeyed to England, where, however, he did not experience a warm reception, his

* Lorenz Theodor Gronovius, a naturalist; author of the 'Museum Ishthyologicum,' published at Leyden in 1754.

innovations in botanical classification having rendered him anything but a *persona grata* to such scientific luminaries as Sir Hans Sloane and Dillenius, the latter being the then professor of botany at Cambridge. In 1737, Linnæus returned to Holland, and published several scientific works; the best known being the 'Genera Plantarum' and the 'Flora Lapponica.' The former of these contained the characters of all the known genera of plants; while the latter was the result of the botanical observations which he had made in his expedition to Lapland. At this time also he paid a visit to Paris, and formed a permanent friendship with Antoine de Jussieu, the first of a famous group of botanists of this name.

In 1738, Linnæus returned to Sweden, having spent three years and a half on his travels. He was now one of the most famous naturalists in Europe, but the reception which he met with from his own countrymen was by no means of a flattering or cordial character; and an attempt to practise as a physician in Stockholm proved at first far from profitable. The university of Göttingen paid him the compliment of offering to him the chair of botany, which he refused; and shortly thereafter one or two lucky hits in his practice brought him into public notice, with the result that he became one of the most popular physicians in Stockholm. He was now offered, and accepted, various scientific and medical preferments, which added not only to his reputation, but also to his income; and towards the close of 1739, he married the daughter of Dr Moræus, to whom he had been so long engaged. In 1741, he was appointed to the professorship of medicine and anatomy in the university of Upsala;

but he shortly afterwards exchanged this for the chair of botany and natural history.

The remainder of the life of Linnæus may be passed over here very briefly. He had now attained the summit of his ambition; and except for the trouble caused him by those almost inevitable controversies which attend the work of all reformers, his life was now free from anxiety and annoyance. For thirty-seven years he remained as professor in Upsala, with the result that this university became the acknowledged centre of botanical and zoological learning in Europe. Students came from all European countries to attend his prelections; and many of his more enthusiastic pupils—such as Hasselquist, Forskal, Solander, and Sparrman—subsequently undertook extensive scientific journeys, by which our knowledge of the fauna and flora of distant regions was largely increased. Linnæus did not allow prosperity to abate his scientific ardour, or to interfere with his scientific labours; and the list of the works which flowed from his pen during this period of his life would be an extremely long one. Of his botanical treatises, the two best known are his 'Philosophia Botanica.' and his 'Species Plantarum.' To zoologists, Linnæus is best known by his 'Systema Naturæ,' which contained a classification of the whole animal kingdom, and which, as has been seen, first saw the light at Leyden in 1735. This classical work went through no less than twelve editions during the lifetime of Linnæus, the last being published in 1766. Linnæus also wrote various medical treatises, none of which, however, would have sufficed to give him enduring fame.

The later years of the life of Linnæus were embittered by domestic annoyances and by failing health. In 1772, he had a slight attack of apoplexy, after which his health rapidly declined; and his remaining years present us, for the most part, with nothing but the melancholy picture of constantly increasing physical decrepitude and mental infirmity. Exhausted with constant suffering, he died on the 10th of January 1778, in the seventy-first year of his age.

Having thus sketched in outline the principal incidents in the life of Linnæus, a brief consideration may be given to the scope and results of his labours as regards natural history.

Like Ray, Linnæus is best known as a botanist; and his fame as a zoologist is principally based upon his classification of the animal kingdom. Ray, as we have seen, adopted a classification of animals which, though greatly in advance of anything which had previously appeared, was nevertheless both artificial and in many respects unnatural. Thus, the Vertebrata were divided into primary groups in accordance with the single character of the structure of the heart; while minor groups were established upon such trivial characters as the possession of an aquatic or terrestrial habit of life. Again, the primary divisions of the Invertebrate Animals were founded upon such an entirely non-essential feature as the mere size of the organism.

The Linnean classification, like that of Ray, was essentially an 'artificial' one, in the sense that the *principle* adopted in framing it was that of selecting some one exclusive character, to which undue importance was attached, and by the possession or absence of which different groups were defined and kept apart. It is to be remembered, however, that though the classification of Linnæus was in principle an artificial one, it was in practice largely natural—that is to say, his groups, though based on artificial distinctions, in many cases really do correspond with natural groups. Moreover, it is to be borne in mind that Linnæus was perfectly well aware of the artificial nature of his system, and that he was acquainted with the requirements of a natural classification. He purposely adopted the 'artificial' principle upon the ground that, in the then state of knowledge, such a classification was alone possible, and could alone be used with advantage. To the Linnean classification, therefore, we must assign the merit of being the most simple and the most complete of all the systems of nature which had been published up to the middle of the eighteenth century; and it not only at once superseded all others, but continued to be in general use for half a century or more after the death of its illustrious author. Under these circumstances no apology is needed for giving here the following brief sketch of the Linnean classification.

In the 'Systema Naturæ,'* Linnæus divided the *imperium naturæ*, or the total assemblage of all natural

---

* The 'Systema Naturæ' bears the full title, 'Systema Naturæ, sive regna tria naturæ systematice proposita per classes, ordines, genera et species.' The first edition consisted of only fourteen folio pages, and was published at Leyden in 1735. The twelfth edition, the last which appeared in the lifetime of Linnæus, was published in three volumes at Stockholm, in 1766-68. The edition which is usually regarded as the authoritative one, is the thirteenth, which was published at Leipsic, in ten volumes, in 1788-93. It was edited by Johann Friedrich Gmelin, who in various ways added to it and amended it. An English translation of this edition, by William Turton, was published in 1802-1806.

objects, into three *regna* or kingdoms—the mineral, vegetable, and animal kingdoms. *Minerals* he defines as bodies or accumulations of matter which have neither life nor sensibility; *vegetables* are organised bodies which have life, but are without sensibility; while *animals* are not only organised and alive, but also possess sensibility and the power of voluntary movement. With Linnæus's arrangement of the mineral kingdom we have nothing to do here. It may be noted, however, that he included in this kingdom all those remains of extinct animals or plants which we now know as 'fossils.' These he placed in a special division of the mineral kingdom (*Larvata*), believing some to be real, while others (such as the Graptolites) were thought by him to be imaginary, that is to say, purely mimetic. We are also not concerned here with the famous Linnean system of plants, though it may be remarked that this constituted an immense advance upon any system of classification of the vegetable kingdom which had preceded it. The animal kingdom was divided by Linnæus into the following six great 'classes.'

(1) *Mammalia, or Quadrupeds*, including the animals which are at the present day placed under this name. Linnæus divided the quadrupeds into seven 'orders,' of which the first, termed *Primates*, included man, the monkeys, and the bats. The characters which induced Linnæus to place the bats with the monkeys are, that in both these groups of animals there are generally only four upper front teeth, and the mammary glands are only two in number and are placed upon the chest. This is a good example of the violation of natural affinities which results

from the selection of some one or two arbitrary characters as the basis of a classification; since there is in reality no close relationship between the bats and the monkeys, and still less between the bats and man. As an instance of the progress of zoological discovery since the time of the great Swedish naturalist, it may be mentioned that while Linnæus, in the twelfth edition of the 'Systema Naturæ,' described or enumerated two hundred and twenty species of quadrupeds, arranged in forty genera, a recent writer (Mr Dobson) is able to describe, in the single order of the Bats alone, some four hundred species, distributed in eighty genera. One of the least satisfactory features in the Linnean classification of the Mammals is, that he broke up the great and natural division of the hoofed quadrupeds, which Ray had established under the name of *Ungulatæ*, and disposed of them by establishing three orders (the *Bruta*, *Pecora*, and *Belluæ*), in which they were placed in unnatural juxtaposition with forms in no way related to them.

(2) *Aves, or Birds.*—Linnæus divided the birds into six orders, comprising less than a thousand species. His classification was a decided advance in clearness and general applicability upon that of Ray and Willughby; and his orders, though to a large extent artificial, have remained in general use, with more or less extensive modifications, even up to the present day.

(3) *Amphibia.*—By this name Linnæus understood the animals which we now know as the Reptiles, such as lizards, turtles, crocodiles, and serpents. At the same time he placed with these the animals which modern naturalists call *Amphibia*, namely, the frogs, toads,

and newts; these in reality being much more nearly allied to the fishes than to the genuine Reptiles. Linnæus also placed in this unnatural class quite a number of true fishes, such as the lampreys, sturgeons, skates, and sharks.

(4) *Pisces, or Fishes.*—Except for the fact that Linnæus, as just remarked, had placed certain fishes among the Reptiles, this class corresponds with what naturalists now understand by the same name. The minor groups of fishes were based upon the presence or absence of the hinder or 'ventral' pair of fins, and the position of these when present, a character of a quite artificial nature; but several of his smaller divisions are very natural. In the department of fishes, however, Linnæus had been preceded by Artedi, and such changes as were made in the 'Systema Naturæ' upon the classification proposed by the latter, were scarcely changes for the better.

(5) *Insecta, or Insects.*—Linnæus was acquainted with about three thousand species of Insects, but he included under this head not only the true Insects, but also the Spiders and Scorpions, the Centipedes, and many of the Crustaceans (crab, lobster, &c.), none of these being, properly speaking, Insects. In other words, the *Insecta* of Linnæus correspond with the 'Articulate Animals' of later writers. Six orders of genuine Insects are recognised, and are defined by the nature of their wings; and a seventh order, under the name of *Aptera*, is retained for the reception of the wingless insects and the other Articulate Animals just mentioned. To a large extent the classification of Insects which was adopted by Linnæus is the one which is now in use; and there is

perhaps no part of his zoological system which has suffered less alteration at the hands of later systematists, that is, as regards its broad features.

(6) *Vermes.*—Under this head, Linnæus included all the Invertebrate Animals, with the exception of the Insects and the other Articulate Animals which he included among the Insects. It need hardly be said, therefore, that the sixth Linnean class is a wholly unnatural and miscellaneous assemblage of animals, most of which do not at all correspond with what we should now call 'Worms.' It was only to be expected that at the time of Linnæus the Vertebrate animals were much better known than the Invertebrate animals, and that the insects should be the best known of all the Invertebrates. Hence, it is no matter of wonder that the Linnean class of *Vermes*—in anything like the Linnean sense—should have totally disappeared in the progress of zoological science, Cuvier having played the most important part in the work of demolition and reconstruction.

Upon the whole, in spite of its defects, and in spite of the fact that it was largely based upon 'artificial' principles, we must concede to the Linnean classification great merits. It was, in fact, as Agassiz has remarked, the first essay, on a large scale, 'at grouping animals together according to certain common structural characters.' It may, however, with some confidence be stated that zoology owed at least as much to Linnæus for the system of scientific nomenclature which he introduced, as it did on the score of his classification of the animal kingdom.

Prior to the time of Linnæus, as has been previously

pointed out, there existed no clear and definite system of zoological nomenclature. No one except a working naturalist can form any conception of the amount of confusion arising from the want of a precise nomenclature. Even with the limited number of specific forms of animals which were known to naturalists in the time of Ray, this confusion was almost intolerably great. At the present day, and in the present state of our knowledge, the study of natural history would be an absolutely hopeless matter, if nomenclature had remained in the condition in which it was at the time of Ray and Willughby. Linnæus, however, has the transcendent merit of having conceived and introduced the so-called 'binomial' system of nomenclature, now in universal use among naturalists. On this system each 'genus,' or group of related 'species,' of animals receives a special Latin name—the 'generic name'—which is used for every species belonging to the group. Each 'species' of the genus is distinguished by a second subordinate title—the 'specific name'—which is placed immediately after the generic name. Consequently, every species of animals is designated by two names, one indicative of the genus to which it belongs, while the other is its own proper appellation. Thus, to give a single example, the dog and the wolf are two 'species' of the 'genus' *Canis*, and they are therefore distinguished from each other as the *Canis familiaris* and the *Canis lupus* respectively. The cat and the tiger, again, are two species of the genus *Felis*, and they therefore stand as the *Felis catus* and the *Felis tigris*.

At the present day, and in certain departments of

natural history, there are indications that it may be necessary to give a further expansion to the Linnean nomenclature, and to adopt a 'trinomial' system, to indicate differences which are slightly below what are usually recognised as 'specific' differences. Even, however, in the case of the general adoption of this modification, the boon conferred upon naturalists by the Linnean system of nomenclature would remain simply beyond calculation.

SIR HANS SLOANE.

THE GREAT MUSEUMS OF BRITAIN.

SIR HANS SLOANE.

AMONG those aids which are indispensable to the general progress of natural history may be reckoned the formation of extensive and well-appointed collections of animals. In this respect few countries occupy a more enviable position than our own, the British Museum and the Hunterian Museum being two of the largest and best ordered collections in existence. It is therefore instructive to glance briefly at the origin of these great and national institutions, each of which owes its inception to the genius of a single individual—in the former case,

Sir Hans Sloane; in the latter instance, the celebrated anatomist and physiologist, John Hunter.

Sir Hans Sloane was born in 1660, and was the youngest son of Alexander Sloane, a Scotchman who had settled in Ireland. In his youth he suffered from ill-health; but few details are known as to his early life. Having determined to adopt the profession of medicine, he went to London, where during four years he diligently studied chemistry, botany, anatomy, physic, and the other subjects which at that time constituted the curriculum for medical students. Thereafter he went abroad, and attended the botanical lectures of Tournefort, a celebrated French botanist, in Paris, subsequently spending a year in the study of the same subject under Professor Magnol of Montpellier. It is believed that he graduated as doctor of medicine at Montpellier; and it is, at any-rate, certain that in 1684 he returned to London, where he settled down to the practice of his profession. Here he started with very brilliant prospects, as he had many influential friends, such as the well-known naturalists, John Ray and Martin Lister, and the famous physician, Dr Sydenham. One evidence of this is the fact that almost immediately after his return to London he was elected a Fellow of the Royal Society, which body, at that time, included almost all the leading scientific workers in the country.

After little more than two years of professional life in London, Sloane was offered the post of physician to the Duke of Albemarle, who was then about to proceed to Jamaica, of which island he had been appointed governor. Sloane recognised in this offer an excellent opportunity of adding to his knowledge of natural history and botany, and unhesitatingly accepted it. Accordingly, in September 1687, he set sail for the West Indies, and, after various halts at different stopping-places, finally reached Jamaica on the 19th of December of the same year. Shortly after his arrival in Jamaica, the Duke of Albemarle died, this untoward event bringing Sloane's official duties to an end before they had well begun; but he remained in the island for rather more than a year, till the arrival of a new governor from England. During this interval, he devoted himself to an examination of the natural productions of Jamaica; and he not only kept a record of his observations, but brought back large and interesting collections. Most of the observations which he made were subsequently given to the world in two extensive treatises— namely, the 'Catalogue of Jamaica Plants' (1696), and the 'Natural History of Jamaica;' the latter being a large and costly folio work, of which the second volume did not appear till as late as 1725.

In the early part of 1689, Sloane returned to London, and again settled down to practice, with the result that he rapidly became one of the most successful physicians of the metropolis. During his long professional career, he held numerous valuable and honourable appointments, amongst which may be mentioned the presidency, for fifteen years, of the College of Physicians. He was also Physician-General to the Army, and he enjoyed the distinction of being the first English medical man upon whom a baronetcy had been conferred (by George I. in 1716).

Not only did Sir Hans Sloane attain great professional

eminence, but he acquired also a high scientific reputation. For nearly twenty years (1693 to 1712) he held the secretaryship of the Royal Society; and for thirteen years (1727-40) he was the president of that learned body. His works on botany and natural history had rendered him well known among naturalists generally; and the wealth which he had acquired in the practice of his profession enabled him not only to keep open house for the benefit of scientific workers from all countries, but also to promote scientific research in various ways. His later life was wholly free from perturbation, or, indeed, eventful occurrences of any kind. When close upon eighty years of age, he retired from active practice, and from public life generally, and resided peacefully in a house which he possessed at Chelsea, where he lived about fourteen years. His friend George Edwards, naturalist and artist, used often to visit him at this period of his life, 'to divert him for an hour or two with the common news of the town, and with anything particular that should happen amongst his acquaintances of the Royal Society;' and to him we owe an interesting picture of Sir Hans Sloane's latest years. He died on the 10th of January 1753, at the age of ninety-three years.

From a scientific point of view, Sir Hans Sloane is best known as the founder of the noble national collection of natural history which is familiar to every one as the British Museum. Two centuries ago, museums had hardly any existence, and there was no *public* collection of natural objects in Great Britain. The first collection of which there is any record in this country was the property of John Tradescant, a native of Holland, who had travelled much abroad, and ultimately settled in London. He founded a well-known botanic garden at Lambeth, and obtained much notoriety through his museum, which was known as 'Tradescant's Ark.' He died somewhere about 1652, and left his collections to his friend Elias Ashmole, a well-known antiquary of his day. By Ashmole the entire collection was ultimately presented to the university of Oxford, where it became the nucleus of the famous 'Ashmolean Museum.' Among the great rarities in Tradescant's museum was a stuffed specimen of the dodo, the great extinct pigeon of the Mauritius, of which the bill and foot are still preserved at Oxford.

Sir Hans Sloane's collections had, as their nucleus and starting-point, the numerous zoological and botanical specimens which he had gathered together during his stay in the West Indies. To these he kept on constantly adding during the whole of his long life. He received by bequest one very extensive collection, which had been got together by Mr William Courten (known later as William Charlton), and which was estimated to have a value of about eight thousand pounds—an enormous sum of money in those days. He also bought for four thousand pounds the collections of Mr James Petiver, who lived in the later part of the seventeenth century. Petiver was a member of the Apothecaries' Company, and a wealthy man; and having many acquaintances among the captains and surgeons of ships, he was able to pick up many rarities. He was, however, more than a mere collector, and not only published several catalogues of portions of his collections, but also contributed to the *Philosophical*

Transactions various papers in different departments of natural history.

By his will, Sir Hans Sloane bequeathed the whole of his vast collections to the nation, upon the single condition that parliament should pay to his family the sum of twenty thousand pounds. Large as this sum was, it was much less than half of what the collections had cost Sloane in actual money, and was believed to be 'not more than the intrinsic value of the gold and silver medals, ores, and precious stones' in the collection. To this offer parliament agreed; and in 1753 an Act was passed for the purchase of Sir Hans Sloane's museum on the above terms. A body of forty-eight trustees was incorporated; Montague House was purchased for the reception of the collections; and the museum was opened to the public on the 15th of January 1759. In this way originated the enormous and wonderful collection of natural objects which is now contained in the magnificent new buildings in Cromwell Road.

JOHN HUNTER.

John Hunter, though best known to the public at large as an eminent surgeon and a great anatomist, possesses claims to immortality of a different kind, but of at least equal validity. He was the first to apply the method of comparative anatomy in a systematic manner to the study of natural history. Indeed, he may almost be said to have been the founder of the science of comparative anatomy, which was afterwards so greatly extended by the labours of the illustrious Cuvier. He was also the founder of the 'Hunterian Museum,' which in its present

JOHN HUNTER.

form is the largest and most complete collection of objects illustrative of comparative anatomy which exists in this country.

John Hunter was the youngest son of a small Lanarkshire farmer, and was named after his father. He was born at Long Calderwood, in the parish of Easter Kilbride, on the 14th of February 1728. His father died when he was only ten years old, and he seems to have been allowed to grow up very much as he liked, with the result that his early education was much neglected, and he never in later life acquired the literary knowledge which he should have obtained when young. At the age of seventeen, he visited Glasgow, in order to assist his brother-in-law, who was a cabinet-maker, and had become embarrassed. It is said—and the statement, if true, is by no means to his discredit—that for a time he actually worked at his brother-in-law's trade; but this has been denied. Eventually, John Hunter returned home, and having no employment there, he determined in 1748 to visit his brother William, who had already obtained a great reputation as a lecturer on anatomy in London, and as to whom a few words may be said here by way of digression.

But for the fact that he has been overshadowed to some degree by the great reputation of his brother John, Dr William Hunter would have bulked more largely in the eyes of posterity than he actually does. He was one of the first anatomists, and admittedly the best anatomical teacher of his day, and was at the same time an eminent and successful physician. He made many important discoveries, most of which are recorded in his 'Medical Commentaries,' and he unfortunately became involved in several warm controversies as to the precise share of credit which he ought to receive for these. This he himself endeavours to excuse upon the ground that 'anatomists have ever been engaged in contention,' and also upon the ground that unless a man resist all 'encroachments upon his discoveries and his reputation, he will hardly ever become considerable in anatomy, or in any other branch of natural knowledge.' It is greatly to be regretted, however, that amongst those with whom he had passages of arms relative to priority of discovery, should have been included his brother John, both of these distinguished men having been of an irascible and irritable temperament. William Hunter accumulated a very extensive museum, mostly of anatomical and physiological preparations, the whole of which he left to the University of Glasgow, in which it constitutes the well-known Hunterian Museum. He was born in 1718, and died in 1783.

As above said, John Hunter finding, when in his twentieth year, that he was still without any profession, determined to visit his brother William, with a view of qualifying himself as a medical man. William Hunter not only consented to receive his brother, but at once placed him as a pupil in his dissecting-room, and also made arrangements for his studying surgery under Cheselden, then the most celebrated surgeon of his day. In anatomy, John Hunter's progress was so rapid, that in 1749 he became demonstrator to his brother, and it was in this capacity that he laid the foundation of that marvellous manipulative skill for which he was in later life so famous. He likewise prosecuted his studies in surgery, which at that time was much less truly scientific than it now is, with the greatest zeal. After acting as assistant to William Hunter for about five years, John was received into partnership with his brother as anatomical lecturer (in 1755). He was, however, at this time, as he continued to be all through his life, but a very indifferent lecturer; whereas his brother William was celebrated for the ease of his delivery and the flowing style of his prelections.

The next four years of John Hunter's life are little more than a mere record of incessant hard work; and the various discoveries which he made at this time are only of interest to specialists. In 1759, he was attacked by inflammation of the lungs, which seems to have left him in an unsatisfactory state of health. His relations with his brother William had also become strained, and it was probably difficult for the two to work together smoothly. With a view, therefore, to change of scene and employment, John applied for an appointment in the army, and was immediately made staff-surgeon. In this capacity he saw a good deal of active service, being present at the siege of Belleisle in 1761. In the year following this, he took part in the campaign against Spain, where he likewise saw a good deal of fighting. One of the fruits of the experience which he gained as military surgeon was the celebrated work, which he published many years afterwards, upon gun-shot wounds.

At the close of the Seven Years' War, Hunter, having recovered his health, again settled down in London, and commenced practice as a surgeon in 1764. At first his progress was not very rapid, partly because his means were small, and partly because his personal manners do not seem to have been such as to quickly win public confidence or professional good-will. Besides practising, Hunter for several years gave lectures on anatomy and operative surgery to a private class. He was at no time, however, a successful lecturer, and his class never amounted to twenty students. At this time of his life, Hunter is said to have been companionable in his habits, and to have mixed freely in society; but his professional engagements began to increase upon him, and he soon became completely absorbed in his scientific pursuits.

More particularly he began now to devote himself to the study of comparative anatomy, a science which had been hitherto little cultivated, and in which he had not only ample scope for his wonderful manipulative skill, but also the certainty of being rewarded by many and important discoveries. At an early period in his career, he had begun to combine with his studies in human anatomy similar investigations into the anatomy of the lower animals; and he had made many preparations illustrative of the structure of the latter. He had also made various interesting discoveries, such as that of the existence of lymphatic vessels in birds; but he did not publish these till a later period. Even during his military service abroad, he had not pretermitted his zoological observations wholly, for we find him varying his duties as army surgeon with experiments on the digestive powers of lizards and snakes at different seasons, and with researches into the auditory powers of fishes. When he returned to London, and settled down as a surgical practitioner and teacher, he renewed his old love for comparative anatomy and physiology; and devoted to this pursuit all his spare time, and, it may be added, all his spare cash. Finding it difficult to carry out many of his investigations at his house in Golden Square, he purchased a piece of ground at Brompton, which was then about two miles out of London, and upon this he built a house, afterwards well known under the name of Earl's Court. Here he used to spend as much of his time as he could spare from his professional avocations, surrounded by all sorts of beasts and birds which had been presented to him, or which he had purchased.

His studies in comparative anatomy and physiology were, however, by no means of purely abstract interest to him, nor disconnected entirely from surgery and medicine. On this point the ablest of his biographers* has made the following apt remarks: 'At the time he [John Hunter] commenced his labours, surgery, though holding a far more respectable station as a practical art than it had done fifty years before, was yet destitute of those sound general views of the nature and treatment of disease, which constitute the foundation of practice in the present day, and the possession of which justly entitles it to claim the rank of a science. The able men who, in this country and on the Continent, immediately preceded Hunter, had succeeded, by the exercise of correct observation and sound judgment, in removing a load of absurd practices with which the art had been clogged; but the improvements suggested by them depended for the most part on isolated experience, and were deficient in a solid and satisfactory foundation upon well-known principles of the animal economy. As yet little had been done towards explaining the real nature of diseases, by showing in what particulars they are allied to natural processes, and what are the aberrations from those processes which give them their peculiar character. Nor were the actions by which nature operates in the cure of diseases at all better understood, and the most vague notions prevailed respecting the important functions of nutrition and absorption, and the processes of adhesion, suppuration, granulation, &c.; the right understanding of which forms as it were the very corner-stone of a good surgical education at the present

* Drewry Ottley.

day. Hunter perceived the want of this knowledge, which in his opinion could alone furnish a sure foundation for the future improvement of surgery; and it was to contribute towards supplying the deficiency that his labours were hereafter to be unceasingly directed.

'He clearly saw, that in order to obtain just conceptions of the nature of those aberrations from healthy actions which constitute disease, it was necessary first to understand well the healthy actions themselves; and that these required to be studied, not in man alone, but throughout the whole animal series, and even to receive further elucidation by comparison with the functions of vegetable life. It was no less an undertaking, then, than the study of the phenomena of life in health and disease, throughout the whole range of organised beings, in which Hunter proposed to engage—an undertaking which required a genius like his to plan, and from the difficulties of executing which, any mind less energetic, less industrious, less devoted to science than his own would have shrunk.'

It should be added that Hunter, in carrying out his scientific researches, adhered rigidly to the sound rule of ascertaining the facts before he framed any theory, rather than of starting with an hypothesis, and seeing how far the facts could be made to square with that. He not only invariably relied upon direct observation and experiment in the collection of his facts, but he showed himself a master of the art of experimental research. He exhibited, indeed, a real genius, not only in the perception of how any particular problem could be solved experimentally, but also in varying his experiments so as to meet the requirements of each special case. In this respect John Hunter has probably rarely had his equal; and it is largely for this reason that his treatment of all the subjects which he took up bore upon it the stamp of originality. 'That man thinks for himself,' is the remark which Lavater is said to have made when shown Hunter's portrait, and assuredly any expert, after reading one of his scientific memoirs, would have arrived at the same conclusion.

In the year 1767, John Hunter was elected a Fellow of the Royal Society, and in the next year he became a member of the Corporation (now the Royal College) of Surgeons. He was also now appointed to the envied post of surgeon to St George's Hospital, an appointment which not only improved his professional status, but also enabled him to obtain private pupils, in accordance with the custom by which the medical students of that time were bound apprentice to some medical man for a term of years. Among those who commenced their medical studies in this way under Hunter's care, was the celebrated Dr Jenner, the discoverer of vaccination; and the relations between these two eminent men, which were to begin with those of master and pupil, ultimately grew into those of a warm friendship, which only ceased with the death of the elder.

In 1771, John Hunter married the eldest daughter of Mr Home, who was surgeon to Burgoyne's regiment of Light Horse, and whose son, subsequently too well known under the name of Sir Everard Home, became ultimately Hunter's assistant, and, in some sort, scientific successor. Hunter's engagement had been, for want of sufficient means, a long one; and it is stated that the expenses which his marriage necessarily entailed upon him were paid for by his well-known treatise entitled 'The Natural History of the Teeth,' the first volume of which he had published some two months previously. Mrs Hunter, according to Ottley, was 'an agreeable, clever, and handsome woman, a little of a *bas-bleu*, and rather fond of gay society, a taste which occasionally interfered with her husband's more philosophic pursuits.'

In spite of his now having taken upon himself the cares of domestic life, and also of his rapidly increasing professional duties, Hunter continued to apply himself with the utmost ardour to the study of comparative anatomy and physiology. Much of his leisure was spent at Earl's Court, where he carried out numerous experiments on digestion, on the growth and destruction of bone, on the metamorphosis of the silkworm, and on many other similar subjects. The following is a short list of some of the more important subjects of a physiological or zoological nature upon which he published papers, with the dates of publication.

In 1772, he published a paper in the 'Philosophical Transactions' on the power which the gastric juice has, under certain circumstances, of 'digesting' or dissolving the walls of the stomach itself after death. This phenomenon, now well known, was new in Hunter's time, and has important bearings upon the chemical theory of digestion.

In 1773, he carried out a dissection of the electric ray or torpedo, and published an account of his observations in the 'Philosophical Transactions.'

In the following year, in the same publication, he

gave an account of the system of cellular cavities or 'air-receptacles' which are connected with the lungs of birds, and which in turn communicate with the interior of many of the bones. He also published a paper on the singular 'Gizzard Trout' or 'Gillarroo Trout' (*Salmo stomachicus*) of Ireland.

In 1775, he published a series of observations on the great electric eel (*Gymnotus electricus*) of South America, and described the anatomical structure of the electrical organs in this fish.

Both in this year and in 1877, he published in the 'Philosophical Transactions' a series of observations on the temperature of animals and plants; and in the latter year he gave to the world the second portion of his 'Treatise on the Natural History of the Teeth.'

Between 1779 and 1785, he published various zoological papers in the 'Philosophical Transactions,' the most important being one on the organ of hearing in fishes.

In 1786, appeared his well-known 'Observations on certain parts of the Animal Œconomy,' in which he not only republished the papers above alluded to, but added various others, dealing with the secretion in the crop of breeding pigeons for the nourishment of their young; on the colour of the *pigmentum nigrum* in different animals; observations tending to show that the wolf, jackal, and dog belong to a single species; on the structure and economy of whales; and various more strictly anatomical or surgical memoirs.

Subsequent to the publication of the 'Animal Œconomy,' Hunter published comparatively few zoological papers, though various observations which he made upon the structure and habits of different animals appeared from time to time, either in the 'Philosophical Transactions,' or in connection with such works as White's 'Journal of a Voyage to New South Wales,' or Russell's 'Natural History of Aleppo.'

Some notion of Hunter's unwearied activity, and of the wide range of the subjects to which he devoted his attention, may be gathered from the constant but intermittently maintained correspondence which he kept up with Jenner. The latter lived in the country, and was in the habit of either supplying Hunter with specimens, or of carrying out experiments under his direction. Unfortunately, we have only Hunter's letters to Jenner left to us, and these are rarely dated; but his requirements from Jenner, and the directions which he gives him, are often most amusing. At one time we find him working at the cuckoo's stomach, and trying to get material for his observations from Jenner. 'I want,' he writes, 'a nest with eggs in it; also one with a young cuckoo; also an old cuckoo.' In the same letter he advises Jenner to remove the egg of a cuckoo into the nest of another bird, and then to tame the young cuckoo and see what sort of a note it has, to which advice he appends the remark—'There is employment for you, young man!' In a later letter, he urges Jenner to 'clear up' the cuckoo, and various other letters contain the same injunction. It may be added that Jenner *did* ultimately communicate to the Royal Society an account of the hatching and rearing of the cuckoo.

Hedgehogs form another subject of endless correspondence between Hunter and Jenner. Hunter dealt with the hedgehog much in the same way as modern physiologists have dealt with the guinea-pig—that is to say, he employed this animal as a convenient vehicle for the carrying out of certain physiological experiments, especially some experiments connected with the temperature of the body. Few of Hunter's letters to Jenner are, therefore, wholly without some allusions to this victim of scientific research; but he seems to have been unlucky as to keeping his hedgehogs alive after he had got them. Thus, in one letter he writes, 'I put three hedgehogs in the garden, and put meat in different places for them to eat as they went along; but they all died.' In another letter, dated a few months later, he writes, 'Have you made any experiments with the hedgehogs, and can you send me some this spring? for all those you sent me died, so that I am hedgehogless.' In a postscript to a later letter, he suggests that Jenner should send him some hedgehogs in 'a box full of holes all round, filled with hay, and some fresh meat put into it;' and Jenner obviously complied with his instructions, for in another letter Hunter acknowledges receipt of the hedgehogs, and adds that one of them was dead on arrival. A year later, he is again appealing to Jenner for more hedgehogs. 'If you could send me a colony of them,' he writes, 'I should be glad, as I have expended all I had except two; one an eagle ate, and a ferret caught the other.'

Among other subjects about which Hunter was constantly writing to Jenner, are such points as the sexes of eels, the spawning of salmon, the migrations of swallows, the anatomy of the porpoise, the temperature of plants, and the nature of fossils. We must not, however, linger longer over this interesting, though unfortunately one-sided correspondence.

In addition to his numerous published observations on comparative anatomy and physiology, Hunter gave to the world various important surgical treatises, one of the most valuable being his work on the blood, inflammation, and gunshot wounds. He was also incessantly engaged in increasing his museum, which may, indeed, be regarded as the great work of his life; but we may leave this subject till we have briefly recounted the chief remaining incidents of his personal life and professional career.

In the main, however, Hunter's life subsequent to his marriage contains little to record beyond what we should expect to find in the case of any successful surgeon in London, who, in addition to his professional duties, should take upon himself the absorbing labours connected with the prosecution of some much-loved non-professional pursuit. Early in 1773, he showed the first symptoms of subsequent disease of the heart, being attacked by a violent paroxysm of *angina pectoris;* but he apparently recovered from this without any impairment of his general health. In the same year, he commenced to lecture publicly on the theory and principles of surgery. He never had a large class, partly because he confined himself almost entirely to the theoretical portions of surgical science, and partly because he remained throughout his life a poor and unattractive public speaker. It has been recorded of him that he never delivered the first lecture of his course without previously taking a dose of opium to dull his sensibilities and abate his nervousness. Owing to the imperfection of his early education, he possessed no faculty of verbal expression, and was unable to trust anything to his memory. According to one of his biographers, 'he wrote his lectures on detached pieces of paper, and such was his confusion that frequently he found himself incapable of explaining his opinion from his notes; and after having in vain attempted to recall the transitory idea, now no longer floating in his mind nor obedient to his will—after having in vain rubbed his face, and shut his eyes, to invite disobedient recollection, he would throw the subject by, and take up another.' A singular contrast, in these respects, were the two brothers, John and William Hunter, the latter having acquired the reputation of being one of the most perfect lecturers and demonstrators that ever lived. Judging, however, from the written copies which we possess of John Hunter's course on surgery, his lectures were strikingly original and suggestive. Moreover, he 'loved truth better than system'—an inestimable virtue in a teacher—and never hesitated to admit a change in his views, if he had been led by further research to alter his previously expressed opinions on any point. It is said, for instance, that Sir Astley Cooper, who was one of his pupils, once exhibited surprise at hearing him express an opinion directly contradictory of something he had said the previous year, and asked him if he had not held quite a different view before. 'Very likely I did,' replied Hunter; 'I hope I grow wiser every year.'

In 1775, Hunter entertained the idea of establishing a metropolitan school of natural history, but the project fell through. In 1776, he received the appointment of surgeon-extraordinary to the king; and in the same year he suffered from a severe attack of some obscure nervous complaint, which laid him up for some months. In 1780, he had an unfortunate controversy with his brother William as to their respective claims to an anatomical discovery of considerable importance, and the result of this was the estrangement for a long period of the two brothers. In 1783, Hunter purchased the remainder of the lease of some extensive premises in Leicester Square, where he erected suitable accommodation for his rapidly extending museum, and whither he removed from his former residence in Jermyn Street.

At this time Mr (after Sir) Everard Home, who was Hunter's brother-in-law, and who had been on foreign service as a staff-surgeon, returned to England, and from this time forward attached himself to John Hunter's fortunes, and became his assistant. Home has left it on record that at this period Hunter was 'at the height of his chirurgical career; his mind and body were both in their full vigour; his hands were capable of performing whatever was suggested by his mind; and his judgment was matured by former experience.' A few years more, however, of almost unbroken prosperity, and of unintermitting labour at his profession, in teaching, and in extending his splendid collection, were all that remained to the great anatomist. In 1785, he began again to suffer from attacks of that most painful and distressing malady, *angina pectoris;* and though he was able to carry on his ordinary avocations, and indeed was able to enjoy existence between the onsets of the disease, he nevertheless from this time onward carried his life in his hands. Any exertion, excitement, or irritation sufficed to bring on a seizure; and as the paroxysms of his malady became more frequent and more severe, even his restless spirit began to recognise the necessity for repose. His brother-in-law now became his assistant in his practice, and later on undertook to deliver his surgical lectures for him. Hunter, however, still continued to publish his researches in various departments of surgery, physiology, or natural history, and still laboured on at his museum. In fact, his practice continued to increase, and his professional duties became more onerous in consequence of his having been appointed, in 1786, Deputy Surgeon-general to the Army, and Inspector-general of Hospitals.

Hunter had long been aware that, to use his own expression, 'his life was in the hands of any rascal who chose to tease and annoy him.' The immediate cause of his death was the excitement consequent on a difference with some of his colleagues at St George's Hospital. At a meeting of the hospital board, one of his colleagues gave a contradiction to some statement which Hunter was at the moment making to the meeting. This excited his passion, and fearing that he might not be able to control his temper, he ceased speaking, and hurried into the adjoining room, where he instantly fell lifeless into the arms of one of the hospital physicians, who happened to be present. All attempts to recall life proved fruitless, and in this way died one of the greatest surgeons and anatomists that Britain has ever produced. He died on the 16th of October 1793, in the sixty-fifth year of his age.

in such a manner as his executors might think best. It was not, however, till 1799 that government could be induced to take the matter up, when parliament voted the sum of fifteen thousand pounds for the purchase of the collection. Having bought the Hunterian collection, government offered it to the Corporation of Surgeons, which in the following year obtained from parliament a royal charter, along with permission to confer diplomas, and which thus became the great corporation now known as the Royal College of Surgeons. The College of Surgeons accepted the charge of Hunter's museum upon the conditions imposed by the government—namely, that they should maintain the efficiency of the collection, throw it open to the Fellows of the college at stated times, prepare a catalogue of it, appoint a conservator, and institute an annual course of lectures on comparative anatomy. In this way arose the present magnificent Hunterian collection of the Royal College of Surgeons, and the hardly less famous Hunterian lectures.*

It was not, however, till 1800 that the College of Surgeons was in a position to take over the collections from Hunter's own museum in Leicester Square, to the temporary buildings allotted for their reception, until such time as a permanent building (for which parliament had voted the further sum of £15,000) could be erected. When the transfer to the temporary buildings had been effected, Sir Everard Home was appointed the first conservator of the museum. At this period Sir Everard Home ordered that all Hunter's manuscripts should be

---

* Two of our most famous living naturalists—namely, Sir Richard Owen and Professor Flower—have been conservators of the Hunterian museum.

removed from the museum to his own private residence, on the ground that they were in need of arrangement. When, however, the trustees of the museum endeavoured to recover these all-important manuscripts, for the purpose of making a catalogue of the museum, they were not forthcoming; and repeated attempts, made year after year, failed to induce Sir Everard to give them up. Ultimately it was ascertained that Sir Everard Home had deliberately burnt the whole of the manuscripts which Hunter had left behind him, upon the ground that Hunter had not mentioned them in his will, and upon the unsupported assertion that Hunter had verbally instructed his brother-in-law to destroy them. Thus were destroyed all Hunter's unpublished observations. It has, however, been explicitly asserted—and there seems to be too good reason to believe that the assertion is correct—that most of these observations *were* published after all, and that they are to be found in the six volumes entitled 'Lectures on Comparative Anatomy' which were subsequently issued by Sir Everard Home, with his own name upon the title-page.

Admirable descriptive catalogues of different portions of the Hunterian collection have been published from time to time during the last fifty years, chiefly through the exertions of Sir Richard Owen. At the present moment, this vast collection, greatly enlarged and enriched as it has been since the death of Hunter, may be regarded as probably the largest and best series of specimens illustrative of comparative anatomy and physiology that has ever been got together beneath a single roof.

BRITISH ZOOLOGISTS.

WHILE the principles of zoological science were being laboriously worked out, and natural history was thus gradually being placed upon a sound and truly philosophical basis, many investigators were engaged in researches upon the animals which inhabited their own countries.

As regards Britain, one of the earliest, as also one of the most meritorious, of the naturalists who have from time to time devoted themselves more particularly to the study of the indigenous animals, was Martin Lister, best known as a conchologist. Lister was born at Radcliffe, in Buckinghamshire, in 1638, and was a contemporary and intimate friend of Ray. Like so many other naturalists, he was by profession a medical man, and he practised for many years in York, subsequently removing to London. He was a well-known physician in his day, having been for a time physician in ordinary to Queen Anne, and he wrote various medical treatises,* which, however, are of no particular value at the present day. His title to fame rests upon his zoological writings, and

---

* The list of his writings, medical and zoological, occupies a column in Watt's *Bibliotheca Britannica*.

principally upon two of these, both dealing mainly with shells—namely, his 'Historia Animalium Angliæ,' and his 'Historiæ sive Synopsis Conchyliorum.' The first of these is, so far as it goes, a 'Natural History of Britain,' but it deals only with the British spiders, the land and freshwater shells, and the marine Mollusca. Added to it is a tract upon British fossils, or, as Lister puts it, 'lapides ad cochlearum imaginem figuratæ.' The second of these is a systematic treatise on shells, and was the best work upon the Mollusca which had appeared up to that time. According to Swainson,* Lister's works on natural history are 'characterised by accurate observation, great knowledge of comparative anatomy, and, in general, just notions of the natural affinities of animals. His various works on shells have laid the foundation of all precise knowledge on this subject;' and Linnæus characterised his history of the Mollusca as the fullest (*ditissimus*) treatise on this group of animals which had appeared up to his time.

Another of the early naturalists who dealt with British animals was the famous Sir Robert Sibbald, best known as an antiquary of no mean pretensions. Sir Robert Sibbald was born in 1641 in Edinburgh, in one of the stormiest periods of the stormy history of Scotland, during the seventeenth century. He adopted the profession of medicine, and studied at Leyden and in Paris, taking his degree at Angers. In 1662 he settled as a physician in Edinburgh, where he practised for some years, and took a prominent part in the establishment of a botanic garden. Having inherited an estate, he left Edinburgh and lived in the country, where he continued to carry

* 'Bibliography of Zoology,' p. 254.

on his scientific pursuits. He was appointed in 1682, by Charles II., Royal Geographer for Scotland—a merely honorary office—and in consequence of this he undertook the preparation of a work on the geography and natural history of his native country. He took a considerable part in the formation and establishment of the College of Physicians of Edinburgh, one result of which was his having the honour of knighthood conferred upon him. Shortly after the accession of James II. to the throne of England, Sir Robert Sibbald followed the example of his friend and patron the Earl of Perth, and turned Roman Catholic. His perversion was so ill received in Scotland, that an attempt was made to assassinate him, and he thought it safest to make his way to London. Here he became ill, and on reflection decided that he had been altogether too precipitate in abjuring his former religious views. He therefore repented of his rashness, and resolved to return home and re-enter the church in which he was born—a laudable resolution which he forthwith carried out.

Sibbald's later life is altogether obscure, and has never been traced out, even the time of his death being unknown, though this event is believed to have taken place about the year 1722. The work by which Sir Robert Sibbald is now best remembered is a large folio volume published in 1684, under the title, 'Scotia Illustrata sive Prodromus Historiæ Naturalis,' &c. The latter half of this work deals with the animals and minerals of Scotland, the descriptions of the animals being accompanied by illustrative plates. Upon the whole, however, Sibbald's name is perhaps most familiar to naturalists in connection with his account of a whale (often referred to as 'Sibbald's whale') which came ashore in November 1690, near Burntisland.

Coming to the eighteenth century, we find many well-known observers devoting themselves, more or less exclusively, to the study of British animals. One of the earliest and most considerable of these was John Ellis, whose name will be permanently associated with the study of Corals and of Zoophytes in general. Very little is known of Ellis's life. He is believed to have been born in Ireland about the year 1710, and he died in 1776. He was a merchant in London, and seems, from occasional allusions in his correspondence, to have experienced the ups and downs often associated with a mercantile life. He early showed a strong taste for the study of natural history in general, and of botany in particular, and contributed many botanical papers to the Royal Society, of which he was a Fellow. His commercial connection with foreign countries, and particularly with the West Indies, led him to write several memoirs dealing with plants having an economic value, the best known of these being his 'Historical Account of Coffee,' published in 1774.

Ellis, however, will be always and best remembered as an investigator into the difficult and at that time little understood group of the corals and their allies. Prior to the researches of Ellis, naturalists had mostly held that corals and 'zoophytes' generally were of a vegetable nature. In fact little doubt was entertained as to the reference of most of these organisms to the vegetable kingdom; though an alternative view was held by some observers, by which all the stony zoophytes were regarded as being mineral productions, and therefore really of an inorganic nature. Ray, for example, unhesitatingly grouped the zoophytes generally among the seaweeds and mosses, though he seems to have thought that some of the harder sorts ('Lithophytes') were perhaps really inorganic. An occasional naturalist, even before the close of the seventeenth century, seems to have ventured to express doubts as to the truth of the prevalent doctrine regarding the vegetable or mineral nature of zoophytes, but the first clear assertion of the animal nature of these organisms was made by Peyssonnel, in a memoir which he laid before the Paris Academy of Sciences in 1727, but which does not seem to have been published. The views of Peyssonnel did not meet with acceptance; and it was not till 1741, when the experiments of Trembley * upon the fresh-water polypes brought the subject again under the notice of the scientific world, that the question was once more seriously discussed. Incited by Trembley's experiments, Bernard de Jussieu investigated the nature of various marine zoophytes carefully, and presented a memoir to the French Academy of Sciences in 1742, in which he maintained the animality of these organisms. The views of Jussieu were accepted

* Abraham Trembley was born at Geneva in 1710, and died in 1784. He is best known through his work dealing with the structure and life-history of the *Hydra* or fresh-water polype ('Mémoires pour servir à l'histoire d'un Genre de Polypes d'eau douce, à bras en forme de cornes,' 2 vols., Paris, 1744). Leuwenhoek had given some account of the *Hydra*, and of its propagating itself by means of buds, as early as 1703; but Trembley first investigated the entire subject of the habits, generation, and power of resisting mutilation of this interesting animal. His experiments were repeated by many naturalists, and an English observer, Henry Baker, published an 'Essay on the Natural History of the Polype,' in which he fully confirmed the chief discoveries which had been announced by the Swiss naturalist.

by the well-known naturalist, Reaumur; but they were in general received with incredulity or entire scepticism.

When the controversy stood at this stage, Ellis was induced to take the matter up, and soon satisfied himself as to the correctness of the views of Peyssonnel and Jussieu. The results of his investigations were from time to time laid before the Royal Society, and were ultimately given to the public in a complete form in his 'Essay towards a Natural History of the Corallines and other Marine Productions of the like kind, commonly found on the Coasts of Great Britain and Ireland.' Not only did Ellis, in this well-known treatise, completely establish the animal nature of the zoophytes in general, but he described and named many species, his descriptions being accompanied by good and for the most part recognisable figures. He also for the first time, showed that the horny urn-shaped capsules ('ovarian vesicles') which are found in the summer months attached to so many of our sea-firs, were really parts of the zoophytes on which they were found. The 'Essay towards a Natural History of the Corallines' was published in London in 1754; it was translated into French in 1756, and became one of the standard treatises upon the group of animals with which it deals.

Ellis's reputation, however, rests largely, not upon the above work alone, but upon a treatise entitled 'The Natural History of many Curious and Uncommon Zoophytes,' published in 1786. This work was a posthumous one, and was based upon a series of plates which Ellis had caused to be drawn, with a view of publishing a general history of zoophytes. The plates were taken from specimens in Ellis's own collection of zoophytes, and scientific descriptions of the figures were added and systematically arranged by the eminent naturalist, Solander, a pupil of Linnæus. The work is therefore always spoken of as 'Ellis and Solander's Natural History of Zoophytes.' The figures of the corals described in this classical treatise are remarkably good; and, to use the words of Lamouroux, 'the beauty, the exactitude, and the perfect execution of the plates placed this work at the head of all those which had up to this time been published.'

The first to write a complete 'British Zoology' was the well-known naturalist and antiquary, Thomas Pennant, one of the most energetic of men, and one of the most voluminous of writers. Pennant was born on the 14th of June 1726, at Downing in Flintshire. He was a descendant of an ancient Welsh family, and as he ultimately inherited the estate of Downing, he was throughout his life an independent and indeed a wealthy man. Very little is known of his personal life, beyond what is revealed incidentally in his writings, or more especially, in that most original and amusing memoir which he published himself under the title of 'My Literary Life.' This, however, is essentially a chronological record of his tours and his principal writings. It is known, however, that he was educated at Oxford; and his taste for natural history was of much earlier date than his going to the university. He himself ascribes his love for the study of nature to the fact that he had been presented, when about twelve years old, with a copy of Willughby's 'Ornithology'—the donor being the father of the celebrated Mrs Piozzi. When he was twenty years old, he made the first of the many journeys or 'tours' which he took in later life. On this occasion he visited Cornwall, where he acquired 'a strong passion for minerals and fossils.' Eight years later he visited Ireland, and, as he invariably did, he kept a journal of his tour, which was extensive enough, embracing, as it did, points as distant as Giants' Causeway on the north, and Cork on the south. 'Owing, however,' as he says himself, 'to the conviviality of the country,' this journal 'never was a dish fit to be offered to the public.'

In the year 1761, Pennant began his great work, the 'British Zoology,' the first edition being a folio, and, when complete, containing one hundred and thirty-two plates. This work went through many editions, in a smaller form, and was translated into Latin and German. The editions most valued are the quarto editions of 1776 and 1777. There is also a good octavo edition of date 1776, in four volumes. The classification adopted in the 'British Zoology' is in the main that of 'the inestimable Ray,' with such alterations as later discoveries seemed to Pennant to render necessary. Pennant had no pretensions to be a comparative anatomist, and his work therefore contains no anatomical details. He gives, however, succinct descriptions of the more conspicuous and easily recognised characters of the animals described in his work, along with an account of their 'uses,' and a history of their habits and mode of life. The figures of the animals described are, further, often fairly characteristic. Pennant was a keen observer, and in many cases the observations which he makes on the habits of animals are not only very interesting but also very accurate.

In other cases, it is difficult to avoid the conclusion that Pennant must have been endowed with a very lively imagination; as, for example, in the circumstantial account which he gives of the 'migrations' of the herring, part of which may be quoted here. According to all our modern knowledge, the herring is a local fish inhabiting the German Ocean, the North Atlantic generally, and the seas north of Asia. The shoals move from place to place, but they are always present in larger or smaller numbers in the seas which they inhabit, and their movements are irregular, and apparently chiefly governed by the conditions affecting food-supply. On the other hand, Pennant, whose account has been referred to by almost all writers on the subject, and was until recently implicitly believed, gives a detailed account of a set annual migration of the herrings to and from the Arctic seas. 'The great winter rendezvous of the herring,' he writes, 'is within the Arctic circle: there they continue for many months in order to recruit themselves after the fatigue of spawning, the seas within that space swarming with insect food to a degree far greater than in our warmer latitudes.

'This mighty army begins to put itself into motion in the spring; we distinguish this vast body by that name, for the word herring is derived from the German *Heer*, an army, to express their numbers.

'They begin to appear off the Shetland Isles in April and May; these are only forerunners of the grand shoal which comes in June, and their appearance is marked by certain signs, by the number of birds, such as gannets and others, which follow to prey on them: but when the main body approaches, its breadth and its depth is such as to alter the appearance of the very ocean. It is divided into distinct columns of five or six miles in length, and three or four in breadth, and they drive the water before them with a kind of rippling: sometimes they sink for the space of ten or fifteen minutes, then rise again to the surface, and in bright weather reflect a variety of splendid colours, like a field of the most precious gems, in which, or rather in a much more valuable light, should this stupendous gift of Providence be considered by the inhabitants of the British Isles.

'The first check this army meets in its march southwards is from the Shetland Isles, which divide it into two parts; one wing takes to the east, the other to the western shores of Great Britain, and fill every bay and creek with their numbers; others pass on towards Yarmouth, the great and ancient mart of herrings; they then pass through the British Channel, and after that in a manner disappear. Those which take to the west, after offering themselves to the Hebrides, where the great stationary fishing is, proceed towards the north of Ireland, where they meet with a second interruption, and are forced to make a second division; the one takes to the western side, and is scarce perceived, being soon lost in the immensity of the Atlantic; but the other, which passes into the Irish Sea, rejoices and feeds the inhabitants of most of the coasts that border on it.'

As an average sample of the style of the 'British Zoology,' we may quote in full the description which Pennant gives of the common hog. After giving the synonymy of the species, and a list of its names in various European languages, he proceeds as follows: 'According to common appearances, the hog is certainly the most impure and filthy of all quadrupeds; we should, however, reflect that filthiness is an idea merely relative to ourselves; but we form a partial judgment from our own sensations, and overlook that wise maxim of Providence, that every part of the creation should have its respective inhabitants. By this œconomy of nature, the earth is never overstocked, nor any part of the creation useless. This observation may be exemplified in the animal before us; the hog alone devouring what is the refuse of all the rest, and contributing not only to remove what would be a nuisance to the human race, but also converting the most nauseous offals into the richest nutriment: for this reason its stomach is capacious, and its gluttony excessive; not that its palate is insensible to the difference of eatables, for where it finds variety, it will reject the worst with as distinguishing a taste as other quadrupeds.

'This animal has (not unaptly) been compared to a miser, who is useless and rapacious in his life, but on his death becomes of public use, by the very effects of his sordid manners. The hog during life renders little service to mankind, except in removing that filth which other animals reject: his more than common brutality urges him to devour even his own offspring. All other domestic quadrupeds show some degree of respect to mankind, and even a sort of tenderness for us in our helpless years; but this animal will devour infants whenever it has opportunity.

'The parts of this animal are finely adapted to its way

of life. As its method of feeding is by turning up the earth with its nose for roots of different kinds; so nature has given it a more prone form than other animals; a strong brawny neck; eyes small, and placed high in the head; a long snout, nose callous and tough, and a quick sense of smelling to trace out its food. Its intestines have a strong resemblance to those of the human species, a circumstance which should mortify our pride. The external form of its body is very unwieldy; yet, by the strength of its tendons, the wild boar (which is only a variety of the common kind) is enabled to fly from the hunters with amazing agility: the back toe on the feet of this animal prevents its slipping while it descends declivities, and must be of singular use when pursued; yet, notwithstanding its powers of motion, it is by nature stupid, inactive, and drowsy; much inclined to increase in fat, which is disposed in a different manner from other animals, and forms a regular coat over the whole body. It is restless at a change of weather, and in certain high winds is so agitated as to run violently, screaming horribly at the same time: it is fond of wallowing in the dirt, either to cool its surfeited body, or to destroy the lice, ticks, and other insects with which it is infested. Its diseases generally arise from intemperance; measles, impostumes, and scrophulous complaints are reckoned among them. *Linnæus* observes that its flesh is wholesome food for athletic constitutions, or those which use much exercise; but bad for such as lead a sedentary life: it is, though, of most universal use, and furnishes numberless materials for epicurism, among which brawn is a kind peculiar to *England*. The flesh of the hog is an article of the first importance to a naval and commercial nation, for it takes salt better than any other kind, and consequently is capable of being preserved longer. The lard is of great use in medicine, being an ingredient in various sorts of plaisters, either pure, or in the form of pomatum; and the bristles are formed into brushes of various kinds.

'This animal has been applied to an use in this island, which seems peculiar to *Minorca* and the part of *Murray* which lies between the *Spey* and *Elgin*. It has been there converted into a beast of draught; for I have been assured by a minister of that county, eye-witness to the fact, that he had on his first coming into his parish seen a cow, a sow, and two *Trogues* (young horses) yoked together, and drawing a plough in a light sandy soil; and that the sow was the best drawer of the four. In *Minorca*, the ass and the hog are common helpmates, and are yoked together in order to turn up the land.

'The wild boar was formerly a native of our country, as appears from the laws of *Hoel Dda*, who permitted his grand-huntsman to chase that animal from the middle of *November* to the beginning of *December*. *William* the Conqueror punished, with the loss of their eyes, any that were convicted of killing the wild boar, the stag, or the roebuck; and *Fitz-Stephen* tells us that the vast forest that in his time grew on the north side of *London*, was the retreat of stags, fallow deer, wild boars, and bulls. *Charles I.* turned out wild boars in the *New Forest, Hampshire*, but they were destroyed in the civil wars.'

The publication of the 'British Zoology' was interrupted in 1765, by a tour which Pennant made on the Continent.

Here he made the personal acquaintance of several of the most distinguished naturalists of the day, amongst whom were Buffon and Pallas.

In 1769, Pennant undertook his first tour in Scotland. As he puts it himself, he had 'the hardiness to venture on a journey to the remotest part of North Britain, a country almost as little known to its southern brethren as Kamschatka.' He was pleased with his visit, and thinks the good report which he gave of the country ought to please the Scotch; as from his having shown that 'it might be visited with safety, it has ever since been *inondée* with southern visitants.' In 1772, Pennant again visited Scotland, penetrating on this occasion as far as the Hebrides. In this tour his success, he says, was equal to his hopes. 'I pointed out everything which I thought would be of service to the country; it was roused to look into its advantages; societies have been formed for the improvement of the fisheries, and for founding of towns in proper places; to all which I sincerely wish the most happy event; vast sums will be flung away; but incidentally, numbers will be benefited, and the passion of patriots tickled.' The journals which he kept of both these tours were published, and contain many interesting observations on the natural history of Scotland, as well as on many other points.

In 1771 was published Pennant's 'Synopsis of Quadrupeds.' This work was republished in 1781, in an enlarged form, under the title, 'History of Quadrupeds.' It went through several editions; and he published a companion work to it under the name of 'The Genera of Birds' (1781).

Pennant seldom allowed a year to pass without making a 'tour' of some kind; and in this way he accomplished very extensive peregrinations through all the principal districts of England, in the Isle of Man, and in parts of Wales. He always kept journals of his travels, which were published, and enjoyed an extensive circulation—a circulation which they well deserved on account of the many observations which they contained as to the antiquities, historical buildings, and natural features of the regions traversed.

The most considerable zoological work which Pennant published, after his 'British Zoology,' is his 'Arctic Zoology.' This well-known work was originally intended to comprehend an account of the natural history of North America; but the author subsequently extended it to include the animals of Northern Europe and Asia. The 'Arctic Zoology' consisted of two quarto volumes, with plates, and was published in 1785. It was translated into German and French; and a second English edition, in three quarto volumes, appeared in 1792.

Only Pennant's principal works have been here alluded to, but his literary activity was incessant, and as varied as it was perennial. No theme was too vast for him, and in his sixty-seventh year he could not only plan an 'Outlines of the Globe' which should extend to fourteen quarto volumes, but he actually possessed energy sufficient to write four of these. The same pen, however, which could write a learned memoir for the 'Philosophical Transactions,' found apparently equally congenial employment in epistles on 'Mail Coaches,' 'Free Thoughts' on the

Militia Laws, or poetical effusions addressed to ladies of his acquaintance.

Those who would form some idea of Pennant's thoroughly original personal character, should read his 'Literary Life,' the only defect of which is that there is not enough of it. The motto to this—*vixi et quem dederat cursum fortuna peregi*—is itself characteristic of the man. His mind, to use his own words, 'was always in a progressive state; it could never stagnate.' At the same time, he was no philosopher or bookworm, but a keen, shrewd, observant man of the world, fond of an active outdoor life, and mixing much in society. 'In the midst of my reigning pursuits,' he says, 'I never neglected the company of my convivial friends, or shunned the society of the gay world.' His energy was extraordinary, as evinced both by his unwearied literary labours, and by the amount of bodily exertion which he underwent. 'Almost all my tours,' he tells us, 'were performed on horseback; to that, and to the perfect ease of mind I enjoyed in these pleasing journeys, I owe my *viridis senectus;* I still retain, as far as possible, the same species of removal from place to place. I consider the absolute resignation of one's person to the luxury of a carriage, to forebode a very short interval between that and the vehicle which is to convey us to our last stage.' In another place he says: 'I am often astonished at the multiplicity of my publications, especially when I reflect on the various duties it has fallen to my lot to discharge—as father of a family, landlord of a small but very numerous tenantry, and a not inactive magistrate. I had a great share of health during the literary part of my days; much of this was owing to the riding exercise of my extensive tours, to my manner of living, and to my temperance. I go to rest at ten; and rise winter and summer at seven, and shave regularly at the same hour, being a true *misopogon*. I avoid the meal of excess, a supper; and my soul rises with vigour to its employs, and (I trust) does not disappoint the end of its Creator.'

This happy, healthy, energetic vitality remained with Pennant almost to the very end of his long life. In his later years his body doubtless 'abated of its wonted vigour;' but his mind still retained 'its powers, its longing after improvements, its wish to receive new light through chinks which time hath made.' When close on seventy years of age, he projected, and energetically commenced his colossal 'Outlines of the Globe;' and though he did not live to carry out this bold conception, he will command universal assent when he says: 'Happy is the old age that could thus beguile its fleeting hours, without injury to any one, and, with the addition of years, continue to rise in its pursuits.' After a comparatively brief period of decay and illness, Pennant passed away on the 16th of December 1798, at the age of seventy-two years. In the long roll of British naturalists he will always hold an honourable place. In the words of Swainson, 'whatever he touched, he beautified, either by the elegance of his diction, the historic illustrations he introduced, or the popular charm he gave to things well known before.'

BRITISH ZOOLOGISTS
(CONTINUED).

DURING the last part of the eighteenth century, and during the few years of the nineteenth century which preceded the appearance of the 'Règne Animal,' natural history was diligently prosecuted in Britain by numerous observers, most of whom can be merely noticed here. For intelligible reasons, the groups of animals most largely studied at this period were birds, fishes, and insects, and to a less extent shellfish (Mollusca). One of the most purely British naturalists of this period was George Montagu, a colonel in the army, and a wealthy man, who left behind him two well-known works on our native animals. One of these is his 'Ornithological Dictionary, or Alphabetical Synopsis of British Birds,' in two octavo volumes, published in 1802. A supplement to this work was published in 1813. The other work was the 'Testacea Britannica, an Account of all the Shells hitherto discovered in Britain,' in two volumes quarto, published in 1803, with a supplement in 1808.

The most extensive writer on ornithology of this period was, however, Dr John Latham, a most voluminous writer, and personally a most estimable man. His three great works are: (1) The 'General Synopsis of Birds,' in eight volumes, small quarto, 1781. (2) The 'Index Ornithologicus,' in two volumes quarto, 1790. (3) 'A General History of Birds,' in eleven quarto volumes, 1821-26. This last is little more than an enlarged edition of the 'General Synopsis.'

Insects have always been a favourite branch of study, and the names of Drury, Smeathman (who gave the first good description of termites), and Moses Harris are familiar to all entomologists. Drury, who was a wealthy jeweller in London, was a great collector of insects, though in no sense himself a naturalist. He sent Smeathman to Africa to collect insects for his cabinet, and he published a work on exotic insects, in which the plates were executed by Moses Harris. This last-named naturalist was an excellent artist, and entomologists still use his 'Aurelian, or Natural History of English Butterflies and Moths.' He also published an 'Exposition of English Insects.'

In general zoology the two most noticeable of the names of this period are Edward Donovan and Dr George Shaw, both of whom were voluminous writers, though neither left any permanent mark in the science of natural history. Donovan's principal works form a series of thirty-eight octavo volumes (1792-1818), dealing respectively with British quadrupeds, British birds, British fishes, British shells, and British insects. He also published 'Illustrations of Entomology, including the Insects of China, India, and New Holland,' in three volumes (1805), and 'The Naturalist's Repository, or Miscellany of Exotic Natural History,' in five volumes (1834). All his works are illustrated by plates which are elaborately and often very beautifully coloured; but the text is of small value. Dr George Shaw, who was assistant-zoologist in the British Museum, was as copious a writer as Donovan, but his works are in the main mere compilations. The two most important are 'General Zoology,' in fourteen volumes (1800-27), and the 'Naturalist's Miscellany,' twenty-four volumes, with more than a thousand plates.

To this period also belongs the celebrated artist-naturalist, Thomas Bewick, so universally famed for his unrivalled delineations of animals, and for the immense advance which he effected in the art of wood-engraving. Thomas Bewick was born at Cherryburn, near Newcastle-on-Tyne, in 1753, and died in 1828. As an artist, his work has received full and critical examination in more than one well-known treatise. As a naturalist, he is best known by his 'General History of British Quadrupeds,' the first edition of which appeared in 1790, and his 'History of British Birds,' of which the first volume appeared in 1797, and the second in 1804. The illustrations of these two works have never been surpassed for power of expression and truth to nature. He possessed 'the royal stamp of genius,' and with it 'the humbler, yet quite as necessary, gift of perseverance; and together these led him to approach nature in simplicity, to receive her lessons with faithfulness, and to depict what he saw with unfailing certainty and loveliness.' *

In addition to the above, the present period produced two naturalists who have obtained a permanent fame for

* 'The Life and Works of Thomas Bewick,' by David Croal Thomson, 1882.

the extraordinary accuracy of their observations, and for their graphic descriptions of the habits of animals—namely, Gilbert White and Alexander Wilson. These two names well deserve something more than a passing mention or a mere enumeration of their published works.

THE REV. GILBERT WHITE.

There are few, probably, to whom the words 'White's Selborne,' do not sound perfectly familiar, though possibly many to whom these words are 'household words' have not actually read the book of which they form the abbreviated title. It must, in truth, be admitted—and the admission cannot be made without some touch of pain and regret—that the press and hurry of the latter half of the nineteenth century render it almost impossible for the majority of people to *read* a book like White's 'Selborne.' Indeed, the art of reading books, in the sense in which our forefathers read them, threatens to become altogether lost; and it is almost inconceivable that a book like White's 'Selborne' should be *written* at the present day. Such books are redolent of the health and peace of the quiet country; they breathe tranquillity and repose; they imply unlimited time for contemplation; they tell of a mind, unresting it may be, but assuredly unhasting. Such ingredients for a book are rarely to be obtained in any age: in the feverish life of modern civilisation they bid fair to disappear altogether.

It has to many appeared a matter for regret that so very little—next to nothing in fact—is known of Gilbert White himself. In this, however, one cannot but feel a sort of fitness, a congruity with the quality of the shy, modest spirit and the tranquil, contented life of the man. Those who would know White, must read White's 'Selborne,' and when they have done that, and learned to love it, they will love the writer of the book, and will know more about him than any biographical enumeration of facts could have told them.

Such facts as *are* known can be stated in very brief compass. Gilbert White was born in Selborne, a little village in the extreme eastern corner of Hampshire, on the 18th of July 1720. His father, John White, who did not at the time of Gilbert's birth reside in Selborne, was the only son of Gilbert White, the vicar of Selborne, and was a barrister in the Middle Temple. When Gilbert was eleven years of age, his father came to reside permanently at Selborne. Little is known of him, but it would seem that Gilbert derived from his father his strong love of nature. He died in 1759, and left instructions in his will that no monument should be erected to him, 'not desiring to have his name recorded, save in the book of life.' Gilbert White was educated at Basingstoke, under the Rev. Thomas Warton, the vicar of Basingstoke. In 1739 he entered Oxford as a student of Oriel College, and he graduated as bachelor of arts in 1743. He must have distinguished himself as a student, since he was elected to a Fellowship of Oriel in 1744, not taking his master's degree till 1746. He seems to have had many opportunities of preferment in the church; but he elected to live peacefully at his old home in his native village, where he died in the seventy-third year of his age, on the 26th of June 1793. Of 'events,' in the ordinary sense of the term, Gilbert White's life possesses none that the most painstaking of biographers has ever been able to discover. He never married, though he was at one time in love. He was of cheerful and sociable disposition, and was beloved by one and all who knew him. He does not appear to have had a large circle of acquaintances; but he maintained a correspondence with several of the leading naturalists of the day, and especially with Pennant.

The work which has rendered Gilbert White immortal is 'The Natural History and Antiquities of Selborne, in the County of Southampton,' the first edition of which was published in London in 1789, in quarto. There have been many subsequent editions, mostly of octavo size. It is not possible to give any notion of this charming book, by abstracting it, by enumerating its contents, or by quotations. Gilbert White was essentially, to use the words of his biographer (Mr Edward Jesse), 'an outdoor naturalist, following the pursuit with unwearied diligence, and enjoying the charms of rural scenery with unbounded admiration.' In his love of nature he resembled Pennant; but the latter was a man of superabundant vitality, wholly unsentimental, self-reliant, self-assertive, and not without a spice of personal vanity; whereas Gilbert White was a serene contemplative soul, devoid of ambition, with the tender, sensitive spirit that a poet—and he wrote poetry occasionally—ought to have. His love of nature was, however, the love of a man of science. In other words, it was a love which was not diminished by close acquaintance with its object, but which, on the contrary, depended on and grew out of the minute and methodical observation of the most trivial details.

'His diaries,' says Mr Jesse, 'were kept with unremitting diligence; and in his annual migrations to Oriel College and other places, his man Thomas, who seems to have been well qualified for the office, recorded the weather journal. The state of the thermometer, barometer, and the variations of the wind are noted, as well as the quantity of rain which fell. We have daily accounts of the weather, whether hot or cold, sunny or cloudy; we have also information of the first tree in leaf, and even of the appearance of the first fungi, and of the plants first in blossom. We are told when mosses vegetate, and when insects first appear and disappear. There are also remarks with regard to fish and other animals; with miscellaneous observations and memoranda on various subjects.' Through this mass of what many people might regard as insignificant, not to say wearisome, details, runs a strong vein of *humanity*. To quote Mr Jesse once more: 'He "chronicled" his ale and beer, as they were brewed by his man Thomas, who appears to have been his valet, gardener, and assistant naturalist. He takes notice of the quantity of port wine which came to his share when he divided a pipe of it with some of his neighbours; and he makes frequent mention of his crops, his fine and early cucumbers, and the flavour of his Cardilliac peas—he evidently passing much time in his garden. The appearance of his neighbours' hops, the beginning and ending of their harvests, their bees, pigs, and poultry, are also noticed in succession, and appear to have added to the interest he took in rural life.'

He had also a whimsical sense of the humour that underlies many of the actions of animals, or many of the phenomena of animal life, when viewed from the human standpoint. Take, for example, the account which he gives of 'Timothy,' a large and aged tortoise, which he kept for many years, and the habits of which he observed with the same loving care as he bestowed upon all living beings. 'The old Sussex tortoise,' he writes to his friend the Hon. Daines Barrington, 'that I have mentioned to you so often, is become my property. I dug it out of its winter dormitory in March last, when it was enough awakened to express its resentments by hissing; and, packing it into a box with earth, carried it eighty miles in post-chaises. The rattle and hurry of the journey so perfectly roused it, that when I turned it out on a border, it walked twice down to the bottom of my garden; however, in the evening, the weather being cold, it buried itself in the loose mould, and continues still concealed. . . . When one reflects on the state of this strange being, it is a matter of wonder to find that Providence should bestow such a profusion of days, such a seeming waste of longevity, on a reptile that appears to relish it so little as to squander more than two-thirds of its existence in a joyless stupor, and be lost to all sensation for months together in the profoundest of slumbers. . . . Because we call this creature an abject reptile, we are too apt to undervalue his abilities, and to depreciate his powers of instinct. Yet he is, as Mr Pope says of his lord,

> Much too wise to walk into a well;

and has so much discernment as not to fall down a haha,

but to stop and withdraw from the brink with the readiest precaution.

'Though he loves warm weather, he avoids the hot sun; because his thick shell, when once heated, would, as the poet says of solid armour, "scald with safety." He therefore spends the more sultry hours under the shelter of a large cabbage leaf, or amidst the waving forests of an asparagus bed. But as he avoids the heat in summer, so, in the decline of the year, he improves the faint autumnal beams by getting within the reflection of a fruit-wall; and though he never has read that planes inclining to the horizon receive a greater share of warmth, he inclines his shell, by tilting it against the wall, to collect and admit every feeble ray.

'Pitiable seems the condition of this poor embarrassed reptile: to be cased in a suit of ponderous armour; to be imprisoned, as it were, within his own shell, must preclude, we should suppose, all activity and disposition for enterprise. Yet there is a season of the year (usually the beginning of June) when his exertions are remarkable. He then walks on tiptoe, and is stirring by five in the morning; and, traversing the garden, examines every wicket and interstice in the fences, through which he will escape if possible; and often has eluded the care of the gardener, and wandered to some distant field. The motives that impel him to undertake these rambles seem to be of the amorous kind. His fancy then becomes intent on sexual attachments, which transport him beyond his usual gravity, and induce him to forget for a time his ordinary solemn deportment.'

As an admirable example of Gilbert White's wonderful

power of minute observation, we may quote his remarks on the motions of birds, though there is hardly a letter in this charming book that does not exhibit the same close and faithful observation of nature in its out-of-door aspects. 'A good ornithologist,' he writes, 'should be able to distinguish birds in the air, as well as by their colours and shape, on the ground as well as on the wing, and in the bush as well as in the hand. For, though it must not be said that every species of birds has a manner peculiar to itself, yet there is somewhat, in most genera at least, that at first sight discriminates them, and enables a judicious observer to pronounce upon them with some certainty. Put a bird in motion,

> Et vera incessu patuit
> (And it is truly declared by its gait).

Thus kites and buzzards sail round in circles, with wings expanded and motionless; and it is from their gliding manner that the former are still called, in the north of England, gleads, from the Saxon verb *glidan*, to glide. The kestrel, or windhover, has a peculiar mode of hanging in the air in one place, his wings all the while being briskly agitated. Hen-harriers fly low over heaths or fields of corn, and beat the ground regularly like a pointer or setting dog. Owls move in a buoyant manner, as if lighter than the air; they seem to want ballast. There is a peculiarity belonging to ravens that must draw the attention even of the most incurious—they spend all their leisure time in striking and cuffing each other on the wing in a kind of playful skirmish; and when they move from one place to another, frequently turn on their backs with a loud croak, and seem to be falling on the ground. When this odd gesture betides them, they are scratching themselves with one foot, and thus lose the centre of gravity. Rooks sometimes dive and tumble in a frolicsome manner; crows and daws swagger in their walk; woodpeckers fly *volatu undosa*, opening and closing their wings at every stroke, and so are always rising and falling in curves. All of this genus use their tails, which incline downwards, as a support while they run up trees. Parrots, like all hooked-clawed birds, walk awkwardly, and make use of their bill as a third foot, climbing and descending with ridiculous caution. All the Gallinæ parade and walk gracefully, and run nimbly; but fly with difficulty, with an impetuous whirring, and in a straight line. Magpies and jays flutter with powerless wings, and make no despatch; herons seem encumbered with too much sail for their light bodies; but these vast hollow wings are necessary in carrying burdens, such as large fishes and the like; pigeons, and particularly the sort called smiters, have a way of clashing their wings, the one against the other, over their backs, with a loud snap; another variety, called tumblers, turn themselves over in the air. Some birds have movements peculiar to the season of love; thus ringdoves, though strong and rapid at other times, yet in the spring, hang about on the wing in a toying and playful manner; thus the cocksnipe, while breeding, forgetting his former flight, fans the air like a windhover; and the greenfinch, in particular, exhibits such languishing and faltering gestures as to appear like a wounded and dying bird. The kingfisher darts along like an arrow; fern-owls, or goatsuckers, glance in the dusk over the tops of trees like a meteor;

starlings, as it were, swim along; while missel-thrushes use a wild and desultory flight; swallows sweep over the surface of the ground and water, and distinguish themselves by rapid turns and quick evolutions; swifts dash round in circles; and the bank-martin moves with frequent vacillations like a butterfly. Most of the small birds fly by jerks, rising and falling as they advance. Most small birds hop; but wagtails and larks walk, moving their legs alternately. Skylarks rise and fall perpendicularly as they sing; woodlarks hang poised in the air; and titlarks rise and fall in large curves, singing in their descent. The whitethroat uses odd jerks and gesticulations over the tops of hedges and bushes. All the duck-kind waddle; divers and auks walk as if fettered, and stand erect on their tails; these are the *compedes* of Linnæus. Geese and cranes, and most wild-fowls, move in figured flights, often changing their position. The secondary remiges* of Tringæ, wild-ducks, and some others, are very long, and give their wings, when in motion, a hooked appearance. Dabchicks, moor-hens, and coots fly erect, with their legs hanging down, and hardly make any despatch; the reason is plain, their wings are placed too far forward out of the true centre of gravity; as the legs of auks and divers are situated too far backward.'

Among the innumerable subjects which Gilbert White investigated, there seem to have been few that interested him more than the problem of the disappearance of the swallows at the approach of winter. On this point he never could make up his mind to fully accept the ordinary theory of the migration of these birds in autumn to some

* Wing-feathers.

warmer region. On the contrary, he clung to the widely-spread belief that 'many of the swallow kind do not depart from this island, but lay themselves up in holes and caverns, and do, insect-like and bat-like, come forth at mild times, and then retire again to their *latebræ* or lurking-places.' To this idea he recurs again and again; and he does not even appear to be quite clear, but that the popular northern notion that swallows in autumn bury themselves in the mud at the bottom of streams and ponds may not have some truth in it. He seems to have corresponded much with Pennant as regards this *quæstio vexata;* and the latter devotes considerable space in his 'British Zoology' to a discussion of the evidence on the point. Pennant's own conclusion was that the greater portion of the swallows migrate to some warmer country, but that the late broods, being unfit for so arduous a journey, hybernate in this country. We are obliged, says Pennant, to conclude that 'one part of the swallow tribe migrate, and that others have their winter quarters near home. If it should be demanded, why swallows alone are found in a torpid state, and not the other many species of soft-billed birds, which likewise disappear about the same time? the following reason may be assigned:

'No birds are so much on the wing as swallows, none fly with such swiftness and rapidity, none are obliged to use such sudden and various evolutions in their flight, none are at such pains to take their prey, and we may add, none exert their voice more incessantly; all these occasion a vast expense of strength and of spirits, and may give such a texture to the blood, that other animals cannot experience; and so dispose, or we may say, necessitate, this tribe of birds, or part of them at least, to a repose more lasting than that of any others.

'The third notion is, even at first sight, too amazing and unnatural to merit mention, if it was not that some of the learned have been credulous enough to deliver, for fact, what has the strongest appearance of impossibility; we mean the relation of swallows passing the winter immersed under ice, at the bottom of lakes, or lodged beneath the water of the sea at the foot of rocks. The first who broached this opinion, was *Olaus Magnus*, Archbishop of *Upsal*, who very gravely informs us, that these birds are often found in clustered masses at the bottom of the northern lakes, mouth to mouth, wing to wing, foot to foot; and that they creep down the reeds in autumn to their subaqueous retreats. That when the old fishermen discover such a mass, they throw it into the water again; but when young inexperienced ones take it, they will, by thawing the birds at a fire, bring them indeed to the use of their wings, which will continue but a very short time, being owing to a premature and forced revival.

'That the good archbishop did not want credulity, in other instances, appears from this, that after having stocked the bottom of the lakes with birds, he stores the clouds with mice, which sometimes fall in plentiful showers on *Norway* and the neighbouring countries.

'Some of our own countrymen have given credit to the submersion of swallows; and Klein patronises the doctrine strongly, giving the following history of their manner of retiring, which he received from some countrymen and others. They asserted that sometimes the swallows assembled in numbers on a reed, till it broke and sunk with them to the bottom; and that their immersion was preluded by a dirge of a quarter of an hour's length. That others would unite in laying hold of a straw with their bills, and so plunge down in society. Others again would form a large mass, by clinging together with their feet, and so commit themselves to the deep.

'Such are the relations given by those that are fond of this opinion, and though delivered without exaggeration, must provoke a smile. They assign not the smallest reason to account for these birds being able to endure so long a submersion without being suffocated, or without decaying, in an element so unnatural to so delicate a bird; when we know that the otter, the cormorant, and the grebes, soon perish if caught under ice, or entangled in nets; and it is well known that those animals will continue much longer under water than any others to whom nature hath denied that particular structure of heart necessary for a long residence beneath that element.'

ALEXANDER WILSON.

Alexander Wilson, 'the American ornithologist,' was the son of a 'wabster,' or weaver, of Paisley, and was born on the 6th of July 1766. Little or nothing is known of his early life, except that his parents cherished the ambition, so common among the Scotch peasantry, of bringing up a son to the church, and that, with this end in view, Alexander Wilson was for a time placed under the charge of Mr Barlas, a student of divinity, from whom we may suppose he acquired some rudiments of

ALEXANDER WILSON.

a liberal education. Much, however, cannot be set down to this, because his father found it impossible to carry on his education, and the future naturalist was apprenticed, at the early age of thirteen, to an operative weaver in Paisley. His apprenticeship lasted three years, and though often, as he says himself, 'feasted wi' the hazel-oil,' he does not seem to have been unhappy, and he was not wholly without opportunities of indulging his taste for reading, and for wandering in the country.

On the expiry of his apprenticeship, Wilson adopted the occupation of a journeyman weaver, which he carried on, partly in Paisley and partly in Lochwinnoch, where his father now lived, for a period of about four years. Subsequently he set up as a weaver in Paisley, in partnership with a Mr David Brodie, afterwards a schoolmaster, with whom he established a firm friendship, and who assisted him greatly in studying both the more famous English poets and some of the Latin classics. After a time, Wilson tired of his occupation as a weaver, and determined to set up as a pedlar, an avocation which would allow him to gratify his love of nature and his taste for wandering, and at the same time would supply him with bread and cheese. He soon found, however, that the itinerant life of a pedlar was by no means free from hardships and annoyances, and he ultimately returned to Paisley, with the intention of trying to utilise his poetical talents. He had always had a taste for poetry, and had committed many of his effusions to paper; and he now submitted these to a gentleman in Paisley, upon whose judgment he relied. The verdict being favourable, he printed a small volume of poems, and again started off as a pedlar; his pack, however, now containing not only the ordinary wares of his profession, but also a number of copies of his book. This novel attempt to combine poetry with hawking proved, as was to be expected, pecuniarily unprofitable; and Wilson returned to Paisley, 'nearly penniless, and much depressed in spirits.'

In consequence of this failure of his scheme for conquering fame and earning a living through his poems, Wilson was obliged once more to take up his old trade of weaving, at anyrate as an occasional thing; and he fell into a low condition both of body and of mind. Ultimately, he once more resumed his pack, and set off again on his wanderings. At the same time, he began to write for the periodicals of the day, and made a successful hit with his well-known ballad of 'Watty and Meg;' but his literary struggles at this period of his life do not concern us here. The west of Scotland was at this time in an extremely unsettled state, partly owing to depression in trade, and partly to the general restlessness produced in the working-classes of almost all countries by the French Revolution. The operatives of the west country, rightly or wrongly, believed themselves to be an oppressed and ill-treated race, and their cause was warmly espoused by Wilson, who wrote a number of poetical squibs, attacking both men and measures. These, unfortunately, brought him under the notice of the authorities, with the result that a prosecution was instituted against him, and he was condemned to imprisonment in Paisley jail, and also to burn one of his obnoxious pieces with his own hands, in public.

On his liberation, Wilson determined to emigrate to America, whither he set sail on the 23d of May 1794, arriving in the state of Delaware some twenty-two days after leaving Belfast. The first four years of Wilson's life in America seem to have been passed in different employments, but next to nothing is known of the details of his life at this time. In 1800, we find him keeping a school at Frankfort, Pennsylvania, and we have also glimpses of him as a land-surveyor, and as a leader in a local debating society. It is, however, clear that he had, upon the whole, been an unsuccessful man, and that his experiences in his new fatherland had not been agreeable; since he is found in 1801 writing to his friend Ord, urging him not to make any engagements which would bind him to the 'unworthy soil' of America, and looking forward to a speedy return to Scotland.

Things were now to go somewhat more smoothly with Wilson, and the beginning of a better time was ushered in by his obtaining a post as schoolmaster at Gray's Ferry, on the Schuylkill River, about four miles from Philadelphia. Here he made acquaintance with William Bartram,* who kept a botanic garden on the western bank of the Schuylkill, and was a good naturalist. Bartram introduced Wilson to the study of natural history, and induced him to use his pencil as a draughtsman of birds and other natural objects. This proved the germ of what soon became the one overmastering passion of his life, and inspired him with the idea of illustrating the ornithology of the United States. He now spent all his spare time in studying natural history, in improving himself as a draughtsman, in adding to his collections, and in making excursions for the purpose of increasing his store of knowledge as to the habits of the birds of America. He even learned, with the help of his friend Mr Lawson, who was himself an engraver, to put his plates on the copper with his own hands. It need hardly be said that the zeal which he displayed in his ornithological pursuits proved by no means beneficial to his school; and the prospect of his being able to carry out his design of preparing a work on the birds of the United States seemed further off than ever. He proposed to his friend Lawson, in 1806, to join him in the production of such a work; but the latter

* William Bartram was the son of John Bartram, M D., a naturalist. He was the author of 'Travels in North and South Carolina,' &c., Philadelphia, 1791.

declined to do so, upon prudential grounds, and Wilson determined to prosecute his plan unaided. At the worst, as he said himself, he should, in so doing, at least leave 'a small beacon,' to point out where he perished.

Wilson now applied to Jefferson, then president of the United States, for some post in connection with a contemplated expedition to explore the region drained by the Mississippi; but, for some unexplained reason, his application received no notice. However, fortune now smiled upon him, and he was rendered to some extent independent by being appointed assistant-editor on 'Rees's Cyclopædia,' a new edition of which was about to be brought out by Mr Samuel F. Bradford, a bookseller of Philadelphia. He did not delay long in submitting to Mr Bradford his plan for preparing an ornithology of the United States; and Mr Bradford not only gave the plan his approval, but agreed to become publisher, and to find the necessary funds.

Wilson was now able to devote himself to his great enterprise, whenever his editorial duties would allow him to do so; and he spent the next two years hard at work on his contemplated treatise. 'At length,' to use the words of his biographer and friend, Mr George Ord, 'in the month of September 1808, the first volume of the 'American Ornithology' made its appearance. From the date of the arrangement with the publisher, a prospectus had been issued, wherein the nature and intended execution of the work were specified. But yet no one appeared to entertain an adequate idea of the treat which was about to be afforded to the lovers of the fine arts and of elegant literature; and when the superb volume was presented to the public, their delight was equalled only by their astonishment that America, as yet in its infancy, should produce an original work in science, which could vie in its essentials with the proudest productions of a similar nature of the European world.'

The remaining years of Wilson's too short life were occupied principally with the preparation of the concluding volumes of his great work. When not busily engaged on this, he was often away, for months together, on long and toilsome excursions, in search of 'birds and subscribers.' In these journeys, he travelled over a large part of the United States, and underwent many hardships. Many of the incidents of his travels are recounted in his letters to his friends, which also contain numerous exceedingly interesting observations on the condition of the United States at that time, the social habits of the people, and other cognate matters. These letters are most interesting reading, but they do not lend themselves to extraction or quotation, and they must be read in their entirety by those who would like to follow the adventurous naturalist in his wanderings.

In the intervals between his excursions, Wilson applied himself unremittingly to the publication of the successive volumes of his work; and in the beginning of 1813 the seventh volume made its appearance. He was in hopes that the eighth volume would complete the undertaking, but he was not destined to see the end of his labours. Too severe application to his self-imposed task had impaired his health, and he was seized with an attack of dysentery, which in a few days terminated fatally. He died on the 23d of August 1813, in the forty-eighth year of his age.

The 'American Ornithology,' by which the name of Alexander Wilson will be ever remembered, consisted of eight volumes quarto, to which a ninth volume was subsequently added by Prince Charles Lucien Bonaparte, Prince of Musignano. The best English edition is that by Sir William Jardine, in three volumes octavo, published in 1832. From the purely zoological point of view, this great work stands in an almost unique position. Wilson was no trained naturalist, and naturally cared little for zoological systems, for synonymy—the bugbear of working naturalists—or for comparative anatomy. On the other hand he was a wonderfully acute and accurate observer, with a keen sense of what constituted the really essential characters of a species, and thoroughly able to set down these characters in clear and well-chosen language. Above all, he was an outdoor naturalist. He did not describe the birds of America from skins or stuffed specimens, but from close personal observation of the creatures themselves in their native haunts. Hence, as Swainson has remarked, 'his descriptions are, in fact, biographies.'

In illustration of the above remarks, we may quote some portions of the description which Wilson gives of the fish-hawk or osprey. 'This formidable, vigorous-winged, and well-known bird,' writes Wilson, ' subsists altogether on the finny tribes that swarm in our bays, creeks, and rivers; procuring his prey by his own active skill and industry; and seeming no further dependent on the land than as a mere resting-place, or, in the usual season, a spot of deposit for his nest, eggs, and young. The figure here given is reduced to one-third of the size of life, to

FISH-HAWK, FROM WILSON'S 'ORNITHOLOGY.'

correspond with that of the bald eagle, his common attendant and constant plunderer.*

'The fish-hawk is migratory, arriving on the coasts of New York and New Jersey about the 21st of March, and retiring to the south about the 22d of September. Heavy equinoctial storms may vary these periods of arrival and departure a few days; but long observation has ascertained that they are kept with remarkable regularity. On the arrival of these birds in the northern parts of the United States, they sometimes find the bays and ponds frozen, and experience a difficulty in procuring fish for many days. Yet there is no instance on record of their attacking birds, or inferior land animals, with intent to feed on them; though their great strength of flight, as well as of feet and claws, would seem to

* This statement applies to the original figure in the American edition of the work. The figure here given is one-eighth of the natural size.

render this no difficult matter. But they no sooner arrive, than they wage war on the bald eagles, as against a horde of robbers and banditti; sometimes succeeding, by force of numbers and perseverance, in driving them from their haunts, but seldom or never attacking them in single combat.

'The first appearance of the fish-hawk in spring is welcomed by the fishermen, as the happy signal of the approach of those vast shoals of herrings, shad, &c. that regularly arrive on our coasts, and enter our rivers in such prodigious multitudes. Two of a trade, it is said, seldom agree; the adage, however, will not hold good in the present case, for such is the respect paid to the fish-hawk, not only by this class of men, but, generally, by the whole neighbourhood where it resides, that a person who should attempt to shoot one of them would stand a fair chance of being insulted. This prepossession in favour of the fish-hawk is honourable to their feelings. They associate with its first appearance ideas of plenty, and all the gaiety of business; they see it active and industrious like themselves, inoffensive to the productions of their farms; building with confidence, and without the least disposition to concealment, in the middle of their fields, and along their fences; and returning, year after year, regularly to its former abode.

The nest of the fish-hawk is usually built on the top of a dead or decaying tree, sometimes not more than fifteen, often upwards of fifty, feet from the ground. It has been remarked by the people of the sea-coasts, that the most thriving tree will die in a few years after being taken possession of by the fish-hawk. This is attributed to the fish-oil, and to the excrements of the bird; but is more probably occasioned by the large heap of wet salt materials of which the nest is usually composed. I ascended to several of these nests that had been built in from year to year, and found them constructed as follows: Externally, large sticks, from half an inch to an inch and a half in diameter, and two or three feet in length, piled to the height of three or four feet, and from two to three feet in breadth; these were intermixed with corn-stalks, sea-weed, pieces of wet turf in large quantities, mullein stalks, and lined with dry sea-grass; the whole forming a mass very observable at half a mile's distance, and large enough to fill a cart, and be no inconsiderable load for a horse .... About the first of May, the female fish-hawk begins to lay her eggs, which are commonly three in number, sometimes only two, and rarely four. They are somewhat larger than those of the common hen, and nearly of the same shape. The ground-colour varies in different eggs, from a reddish cream to nearly white, splashed and daubed all over with dark Spanish brown, as if done by art. During the time the female is sitting, the male frequently supplies her with fish; though she occasionally takes a short circuit to sea herself, but quickly returns again. The attention of the male, on such occasions, is regulated by the circumstances of the case. A pair of these birds, on the south side of the Great Egg Harbour River, near its mouth, was noted for several years. The female, having but one leg, was regularly furnished, while sitting, with fish in such abundance, that she seldom left the nest, and never to seek for food. This kindness was continued both before and after incubation. Some animals, who claim the name and rationality of man, might blush at the recital of this fact. On the appearance of the young, which is usually about the last of June, the zeal and watchfulness of the parents are extreme. They stand guard, and go off to fish, alternately; one parent being always within a short distance of the nest. On the near approach of any person, the hawk utters a plaintive whistling note, which becomes shriller as she takes to wing, and sails around, sometimes making a rapid descent, as if aiming directly for you; but checking her course, and sweeping past, at a short distance overhead, her wings making a loud whizzing in the air. . . . . It is universally asserted by the people of the neighbourhood where these birds breed, that the young remain so long before they fly, that the parents are obliged at last to compel them to shift for themselves, beating them with their wings, and driving them from the nest. But that they continue to assist them even after this, I know to be a fact, from my own observation, as I have seen the young bird meet its parent in the air, and receive from him the fish he carried in his claws.

'The flight of the fish-hawk, his manœuvres while in search of fish, and his manner of seizing his prey, are deserving of particular notice. In leaving the nest, he usually flies direct till he comes to the sea, then sails round, in easy curving lines, turning sometimes in the air as on a pivot, apparently without the least exertion, rarely moving the wings, the legs extended behind, and his remarkable length, and curvature or bend of wing, distinguishing him from all other hawks. The height at which he thus elegantly glides is various, from one hundred to one hundred and fifty or two hundred feet, sometimes much higher, all the while calmly reconnoitring the face of the deep below. Suddenly he is seen to check his course, as if struck by a particular object, which he seems to survey for a few moments with such steadiness, that he appears fixed in air, flapping his wings. This object, however, he abandons; or rather the fish he had in his eye has disappeared, and he is again seen sailing around as before. Now his attention is again arrested, and he descends with great rapidity; but ere he reaches the surface, shoots off on another course, as if ashamed that a second victim had escaped him. He now sails at a short height above the surface, and by a zigzag descent, and without seeming to dip his feet in the water, seizes a fish, which, after carrying a short distance, he probably drops, or yields up to the bald eagle; and again ascends, by easy spiral circles, to the higher circles of the air, where he glides about in all the ease and majesty of his species. At once, from this sublime aerial height, he descends like a perpendicular torrent, plunging into the sea with a loud rushing sound, and with the certainty of a rifle. In a few moments he emerges, bearing in his claws his struggling prey, which he always carries head foremost, and, having risen a few feet above the surface, shakes himself as a water-spaniel would do, and directs his heavy and laborious course directly for the land. If the wind blow hard, and his nest lie in the quarter from whence it comes, it is amusing to observe with what judgment and exertion he beats to windward, not in a direct line, that is, *in the wind's eye*, but making several successive tacks to gain his purpose. This will appear the more striking, when we consider the size of the fish which he sometimes bears along. A shad was taken from a fish-hawk near Great Egg Harbour, on which he had begun to regale himself, and had already eaten a considerable portion of it; the remainder weighed six pounds. Another fish-hawk was passing Mr Beasley's, at the same place, with a large flounder in his grasp, which struggled and shook him so, that he dropt it on the shore. The flounder was picked up, and served the whole family for dinner. It is singular that the hawk never descends to pick up a fish which he happens to drop, either on the land or in the water. There is a kind of abstemious dignity in this habit of the hawk, superior to the gluttonous voracity displayed by most other birds of prey, and particularly by the bald eagle, whose piratical robberies committed on the present species have been already fully detailed in treating of his history.* The hawk, however, in his fishing pursuits, sometimes mistakes his mark, or overrates his strength, by striking fish too large and powerful for him to manage, by whom he is suddenly dragged under; and though he sometimes succeeds in extricating himself, after being taken three or four times down, yet oftener both parties perish. The bodies of sturgeon and several other large fish, with that of a fish-hawk fast grappled in them, have at different

---

* Wilson gives a most graphic and animated account of the habit of the bald eagle of watching till the fish-hawk has caught a fish, and then pursuing him till the latter, 'with a sudden scream, probably of despair and honest execration,' is compelled to drop his prey, which the eagle instantly seizes and carries off.

times been found dead on the shore, cast up by the waves.'

To the above excerpt from Wilson's account of the osprey, may be appended the lines which he wrote on this bird, which allude specially to the friendly feelings with which it is regarded by the fishermen on the Atlantic coast of the United States:

> The osprey sails above the sound,
> The geese are gone, the gulls are flying;
> The herring-shoals swarm thick around,
> The nets are launched, the boats are flying;
>   Yo, ho, my hearts! let's seek the deep,
>   Raise high the song, and cheerly wish her,
>   Still as the bending net we sweep,
>   'God bless the fish-hawk and the fisher!'
>
> She brings us fish, she brings us spring,
> Good times, fair weather, warmth and plenty,
> Fine store of shad, trout, herring, ling,
> Sheepshead and drum, and old-wives dainty,
>   Yo, ho, my hearts! let's seek the deep,
>   Ply every oar, and cheerly wish her,
>   Still as the bending net we sweep,
>   'God bless the fish-hawk and the fisher!'
>
> She rears her young on yonder tree,
> She leaves her faithful mate to mind 'em,
> Like us, for fish, she sails to sea,
> And, plunging, shows us where to find 'em.
>   Yo, ho, my hearts! let's seek the deep,
>   Ply every oar, and cheerly wish her,
>   While the slow bending net we sweep,
>   'God bless the fish-hawk and the fisher!'

CUVIER.

WE must once more turn our attention to the Continent, in order to acquire a notion of the enormous impulse given to natural history, and the vast improvements effected in scientific zoology by the genius of the illustrious Cuvier. Georges Cuvier was born on the 23d of August 1769, at Montbéliard, in the department of Doubs, then belonging to Würtemberg. His father was a retired military man, and was descended from a Protestant family, which had been forced to emigrate from the Jura by the persecutions directed against the Huguenots. Georges Cuvier was a delicate and studious child, and early showed a marked predilection for natural history. When fourteen years old, he was placed at the academy of Stuttgart—the school of Schiller and of other well-known men—and after a brilliant career as a student, he entered the world to seek for his living, at the age of eighteen. A short space of time subsequent to his leaving the academy of Stuttgart was spent as 'sous-lieutenant' in the Swiss regiment of Châteauvieux; but this corps being disbanded, and his family being unable to give him any pecuniary assistance, he accepted the position of tutor in the family of the Comte d'Héricy, who

CUVIER.

... ... Continent, in
... ... supposed given
... ... effected in
... ... of ... ... Cuvier,
... ... of ... ust 1769, at
... ... belonging
... ... tired military, in a...
... ... family, which had
... the ... by the persecution
... ... Georges Cuvier was a
... ... early showed a marked
... When ... teen years old
... of Stuttgart, the school of
... ... after a brilliant
... ... to seek for his
... ...
... ... of Stuttgart was spent
... Swiss regiment of Châteauvieux
... ded, and his family being unable
... assistance, he accepted the posi-
... of the Comte d'H... ... who

CUVIER.

resided near Caen in Normandy. Here was spent a further period of nearly seven years—from 1788 to the end of 1794—in which Cuvier peacefully discharged his tutorial duties, and occupied his leisure time with studying such animals, and particularly such of the lower forms of life, as were accessible to him in his country-retreat. It was in this haven of refuge that Cuvier weathered the stormy period of the 'Reign of Terror,' and it was here that he met the friend who subsequently introduced him to the scientific world of Paris.

Being one night at a meeting of a local agricultural society of which he was secretary, Cuvier was struck with the extraordinary mastery of the subject under debate that evening shown by one of the members of the society. So remarkable was this mastery, that Cuvier at once concluded that the speaker must be the author of the article 'Agriculture' in the 'Encyclopédie Methodique,' and at the end of the speech saluted him as such. Cuvier was right in his conjecture, and the Abbé Tessier, the writer in question, was at first greatly terrified at the discovery of his personality, for he had been sheltering himself at Fécamp from the fury of the Revolution. Cuvier, however, assuaged his fears, and they became fast allies, one result of their friendship being that Tessier wrote strongly to his friends in Paris, recommending Cuvier for some scientific post. In this way Cuvier was introduced to the notice of the well-known naturalists, Lacépède and Geoffroy St Hilaire, who strongly urged him to come to Paris, an invitation which he accepted in the year 1795. Shortly after his arrival in the metropolis, he was appointed professor of natural history in the Central

School of the Pantheon, and in the same year he was elected assistant to Mertrud, the aged incumbent of the new chair of comparative anatomy at the Musée d'Histoire Naturelle.

Cuvier was now fairly launched upon that course of incessant scientific activity which only terminated with his life. In the year of his arrival at Paris (1795) he not only opened his first course of lectures at the Jardin des Plantes, but he published a number of researches on various departments of natural history, such as the structure of the lower larynx in birds, the anatomy of the Roman Snail (*Helix pomatia*), the circulation of the Invertebrata, the structure and classification of the Mollusca (always a favourite study), and the classification of the Invertebrata generally. In the course of the next year Cuvier was elected a member of the newly-founded National Institute, and was associated with Lacépède and Daubenton as the nucleus of the section 'Zoology.' His scientific activity suffered no abatement, and in this year he published a further series of memoirs, of which three are particularly interesting, as showing the first beginnings of his palæontological labours. One of the papers in question dealt with the skeleton of one of the huge extinct American ground-sloths, the Megalonyx, which had previously been regarded as a carnivorous animal, but which Cuvier showed to be truly a gigantic relative of the existing sloths of South America. Another memoir treated of the Megatherium, another extinct ground-sloth; and the third was concerned with the skulls of the cave-bear, which had been found in the famous cavern of Gailenreuth.

In the years 1797 and 1798, Cuvier continued to enrich the science of natural history by a long series of memoirs. In the former of these years we find him writing upon such diverse subjects as the nutritive processes in insects, the structure of the Ascidian Molluscs, the anatomy of the bivalve shell-fish, the nostrils of the Cetaceans, and the characters of the different species of rhinoceros. In the second of these years he was offered the opportunity of accompanying the expedition which Napoleon was to lead to Egypt, but he declined the offer. In this year he not only published his first memoir upon the wonderful series of fossil bones which are found in the Tertiary rocks ('Gypseous Series') of Montmartre, near Paris; but he also gave to the world the first of his separate works—namely, the 'Tableau élementaire de l'Histoire naturelle des Animaux.' This volume contained, in an abbreviated form, the course of lectures which he delivered at the 'École Centrale du Pantheon;' and it may be regarded as the first general statement of the classification of the animal kingdom which he subsequently elaborated more fully, and which, in its main outlines, is still the accepted arrangement of animals.

In the year 1799 Daubenton died, and the chair of natural history in the Collège de France was thus rendered vacant; Cuvier being appointed his successor in 1800. In the latter year he was also appointed secretary of the class of physical and mathematical sciences in the National Institute, an appointment which was made perpetual in 1803, and which he held to the time of his death. In this year he published a number of memoirs, dealing for the most part with the bones of extinct animals. In this year also appeared the first two volumes of his 'Leçons d'Anatomie comparée,' which at once took place as a standard treatise on the subject with which it dealt. In the first two volumes of this work Cuvier was assisted by Dumeril, and in the three later ones by Duvernoy.

In 1801 his principal contributions to science were a memoir on a new species of fossil crocodile, and a second treating of the teeth of fishes. In 1802, Cuvier was appointed commissary of the Institute to accompany the inspectors-general of public instruction, in which capacity he spent some time in the south of France, superintending the foundation of the colleges of Marseilles and Bordeaux. This office he resigned in 1803, in which year he married the widow of M. Duvaucel, a contractor for the public taxes, by whom he had four children, all of whom predeceased him. The next few years of Cuvier's life are noticeable only for their extraordinary fruitfulness in scientific work, each year producing a harvest of valuable memoirs or extensive scientific treatises, which will be noticed immediately. In 1808, he was placed by Napoleon on the council of the Imperial University, in which capacity he on three occasions (1809, 1811, and 1813) presided over commissions charged with the duty of reporting upon the higher educational establishments in those provinces beyond the Alps and the Rhine which had been annexed to France, with the view of affiliating these to the central university of Paris. In his official capacity of perpetual secretary to the Institute, he was further called upon to write the *éloges historiques* on deceased members of the Institute. It was in the same capacity that he was intrusted with the task of drawing up annual reports on the progress of the natural and physical sciences; and it was as perpetual secretary that he prepared his famous 'Rapport historique sur le Progrès des Sciences physiques depuis 1789,' which was published in 1810.

The last twenty years of Cuvier's life were spent under a burden of intellectual labour such as few men could have borne. Not only was his activity as a teacher and as an investigator unabated, but he had now various high administrative posts placed in his hands, the duties of which were both numerous and heavy. He was now, by universal consent, the first of living naturalists, and the scientific honours which were conferred on him are too numerous to mention. He also now became a high state official. Before the fall of Napoleon (1814), he was admitted to the council of state; and Louis XVIII. confirmed him in this office on being restored to the throne of France. In the same year he was elected Chancellor of the University of Paris, and later, he was appointed Grand Master of the Faculties of Protestant Theology, he being himself a Lutheran. In 1819 he was appointed President of the Comité de l'Interieur. In 1824, he was made Grand Officer of the Légion d'Honneur; and in the beginning of 1832 he was raised by Louis Philippe to the rank of peer of France, and subsequently nominated President of the entire Council of State. On the 8th of May he lectured, as it proved for the last time, at the College of France, but he was thereafter attacked by an illness, which commenced in paralysis of the throat, and ultimately affected the respiratory organs. Remedial measures proved of no avail, and on the 13th of

May Cuvier expired, apparently without pain, and retaining his faculties almost to the last breath.

Cuvier's scientific activity between 1802 and 1832 was something extraordinary, especially when his numerous administrative and educational duties are taken into account. Nothing but the briefest possible sketch of his published scientific work during the thirty years in question can be attempted here. Let us first deal with his work as a specialist in zoology. Subsequent to 1802, Cuvier may be considered as having more especially devoted himself* 'to three lines of inquiry, one dealing with the structure and classification of the Mollusca; a second treating of the comparative anatomy and systematic arrangement of the fishes; and the third concerned with fossil Mammals and reptiles primarily, and secondarily with the osteology of living forms belonging to the same groups. As regards the first of these fields of investigation, Cuvier published a long series of papers on the Mollusca, which began as early as 1792, and dealt with almost all the groups now admitted into this sub-kingdom, with the exception of the Polyzoa. Most of these memoirs were published in the "Annales du Museum," between 1802 and 1815, and they were subsequently collected into the well-known and invaluable "Mémoires pour servir à l'Histoire et à l'Anatomie des Mollusques," published in one volume at Paris in 1817. In the department of fishes, Cuvier's researches, begun in 1801, finally culminated in the publication of the "Histoire Naturelle des Poissons." This magnificent work contained descriptions of five

---

* This sketch of Cuvier's scientific work from 1802 to 1817 is taken from an article by the present writer in the last edition of the 'Encyclopædia Britannica.'

thousand species of fishes, and was the joint production of Cuvier and Valenciennes, its publication (so far as the former was concerned) extending over the years 1828-31. Palæontology was always a favourite study with Cuvier, and the department of it dealing with the Mammalia may be said to have been essentially created and established by him. In this region of investigation he published a long list of memoirs, partly relating to the bones of extinct animals, and partly detailing the results of observations on the skeletons of living animals specially examined with a view of throwing light upon the structure and affinities of the fossil forms. In the second category must be placed a number of papers relating to the osteology of the *Rhinoceros indicus*, the tapir, the *Hyrax capensis*, the hippopotamus, the sloths, the manatee, &c. In the former category must be classed an even greater number of memoirs, dealing with the extinct Mammals of the Eocene beds of Montmartre, the fossil species of hippopotamus, the *Didelphys gypsorum*, the Megalonyx, the Megatherium, the cave-hyæna, the extinct species of rhinoceros, the cave-bear, the Mastodon, the extinct species of elephant, fossil species of manatee and seals, fossil forms of crocodilians, chelonians, fishes, birds, &c. The results of Cuvier's principal palæontological and geological investigations were ultimately given to the world in the form of two separate works. One of these is the celebrated "Recherches sur les Ossemens fossiles," in four volumes quarto, published in Paris in 1812, with subsequent editions in 1821 and 1825; and the other is his "Discours sur les Révolutions de la surface du Globe," in one volume octavo, published in Paris in 1825.' This latter work was in reality a second edition, in an enlarged form, of the 'preliminary discourse' to the first edition of the 'Ossemens fossiles.' It was translated into English, and in this form went through several editions. The 'Recherches sur les Ossemens fossiles' is one of the standard works of zoologists and palæontologists, and is likely long to remain so. The edition most commonly used is the fourth, in eight volumes octavo and two quarto volumes of plates, published subsequently to Cuvier's death, and having an introduction by Cuvier's brother Frederick, himself a famous naturalist. The plates were not only in many cases drawn by Cuvier himself, who was an admirable draughtsman, but many of them were also engraved with his own hand.

Famous as are the works just alluded to, nothing that Cuvier published attained a higher reputation, or has been more widely used, than his great systematic treatise on the animal kingdom ('Règne Animal distribué d'après son Organisation'). The first edition of this, in four octavo volumes, appeared in 1817, but in subsequent editions it was much enlarged. 'In this classical work, Cuvier embodied the results of the whole of his previous researches on the structure of living and fossil animals, as giving confirmation and fixity to that system of classification of which he was the originator, and the main features of which still subsist. The whole of this work was his own, with the exception of the Insects, in which he was assisted by his friend Latreille.'

In taking a general view of the advances which Cuvier effected in zoological science, the main results which he accomplished fall naturally under three heads. In the first place, with regard to systematic zoology, and especially to classification, it is necessary, to begin with, to call to mind the condition in which the classification of the animal kingdom had been left by Linnæus, and in which, with unimportant changes, it had remained ever since. Linnæus divided the animal kingdom into the two primary divisions of the 'Animals with red blood,' and 'Animals with white blood,' these divisions corresponding with what Lamarck named the 'Vertebrate Animals' and the 'Invertebrate Animals.' The first of these primary divisions was further split up into the four classes of the Mammalia (quadrupeds) birds, reptiles, and fishes; while the second division was separated into the two classes of the Insects and the Worms (Vermes). Even among the Vertebrate animals, there were features in the Linnean classification of the most unnatural character. Thus, the Cetaceans (whales and dolphins) found themselves with the fishes; while certain of the latter (for example, the sharks) were placed among the reptiles. Amongst the Invertebrate animals, on the other hand, the confusion which reigned was much more extensive. The only definite 'class' of Invertebrates which Linnæus clearly recognised was that of the insects; but he included under this name not only the animals properly known as 'insects,' but also the Crustaceans, the Spiders, the Centipedes, and the Ringed Worms. Thus the Linnean 'Insecta' corresponded, in a general way, with that great division of animals now known by the name of Annulose Animals. On the other hand, all the other Invertebrate animals were placed by Linnæus in a second 'class,' which he termed Vermes. The Vermes of Linnæus constituted, however, an entirely miscellaneous assemblage of animals, bound together by no recognisable bond, and comprising groups having no real affinity with each other.

The Invertebrate animals had early been a favourite study with Cuvier, and he soon recognised the unnatural collocation of the groups composing the Linnean Vermes. As early as 1795, therefore, he came to the conclusion that the Invertebrate animals could be divided into six 'classes' —namely, the Mollusca, the Crustacea, the Worms, the Insects, the Echinoderms, and the Zoophytes. Further researches convinced him of two other points—namely, that certain of these groups were of a rank higher than that assigned to 'classes,' and secondly, that certain of them were so closely related together as to be properly referable to a single division. Ultimately, therefore, Cuvier divided the entire series of the Invertebrate animals into three great 'embranchements,' or, as we should now say, 'sub-kingdoms,' to which he gave the names of the Mollusca, the Articulata, and the Radiata. In the sub-kingdom of the Mollusca he placed the animals which we now know as 'Molluscs,' the barnacles and acorn-shells (Cirripedes) being, however, erroneously included among the true shellfish. The Articulata of Cuvier comprised the Ringed Worms (Annelides), the Crustacea (lobster, crab, &c.), the spiders and scorpions (Arachnida), and the true insects (Insecta). Lastly, in the division of the Radiata, or 'Radiated Animals,' Cuvier included five large groups—namely (1) The Echinoderms (sea-urchins and star-fishes); (2) The Intestinal or Parasitic Worms; (3) The Jelly-fishes (Acalephæ); (4) The Corals and allied animals (Polypi); and (5) The Infusoria, comprising certain microscopic animals. The Vertebrate animals were, finally, raised by Cuvier from the rank of a 'class' to that of a fourth 'embranchement,' equivalent in zoological value to the three primary divisions of the Invertebrate animals.

There can be no question as to the enormous advance presented by the Cuvierian classification of animals, as above sketched out, upon the system established by Linnæus. This advance is seen not only in the greater naturalness of the groups adopted by Cuvier, but also in the fact that his divisions were based upon sounder and more philosophical principles. So far as concerns the actual groups established by Cuvier, naturalists at present unanimously accept the three great divisions, or sub-kingdoms, of the *Vertebrata*, the *Mollusca*, and the *Articulata;* though certain minor changes have been effected in all of these. Thus, the frogs and newts (Amphibia) are now separated from the reptiles proper, with which they were associated by Cuvier, and are regarded as a separate 'class' of Vertebrate animals. Again, the barnacles and acorn-shells (Cirripedes) are now known not to belong to the Molluscs, but to be properly referable to the Crustaceans, and to be therefore 'Articulate animals.' Lastly, the Articulata of Cuvier constitute the 'Annulose animals' of modern zoologists; and it is usual to associate with these the group of parasitic worms, which Cuvier had placed in his lowest 'embranchement' of the animal kingdom.

On the other hand, the Radiata of Cuvier stand to the Cuvierian classification in the same relation that the Vermes did to the Linnean system. In each case, the great systematist, finding himself confronted with a large series of the lower animals with which his acquaintance, as that of all naturalists of the day, was comparatively imperfect, grouped these together into a single great division, which was necessarily ill defined and imperfectly characterised. To Cuvier is due the recognition of the Articulate Animals as a great primary division, and also the separation of the whole series of the Molluscs from the Linnean Vermes, and their establishment as a second great primary division of Invertebrates. It was left to later naturalists to show that the Cuvierian Radiata was really a miscellaneous and artificial group; and that it could be split up into three great divisions, which may be regarded as equivalent to 'sub-kingdoms'—namely, (1) the Echinoderms (sea-urchins and star-fishes); (2) the Cœlenterate animals (corals, sea-anemones, jelly-fishes, and zoophytes generally; and (3) the Protozoa (Infusorian animalcules, &c.).

Not only was the grouping of the animal kingdom adopted by Cuvier much more in accordance with the true relationships of animals than that of Linnæus, but the principles upon which this grouping was based were also much more philosophical. Linnæus, as has been seen, in framing his classification, was principally anxious to supply naturalists with a kind of index to the animal kingdom, so constructed that it might be easy to determine the place in his system of any given animal. He therefore based his arrangement upon the presence or absence of certain easily recognisable, for the most part externally visible, characters. Such external and arbitrary characters, chosen principally because they are easy to recognise, tend, however, as pointed out before, 'to the association of very differently organised species, and as often separate into very remote groups of an artificial system two animals which may have very similar anatomical structures.' *

Cuvier, on the other hand, recognised that the basis of a 'natural' classification could only be found in the anatomy of the animals to be classified. He saw that it was necessary to compare animals with one another, not merely as to the possession of some one particular character—possibly a character in itself of little real importance—but as regards their *whole organisation*. In this way, by the comparison of each animal in all the points of its structure with every other animal, it was possible to show that certain forms agreed with others as to the 'plan' of their organisation. In other words, Cuvier established the fact that certain groups of animals could be shown to be built upon the same general *plan*, irrespective of any modifications which such a plan might undergo; whereas other groups of animals were built upon a different fundamental plan. Moreover, Cuvier showed that in comparing animals with one another it was necessary to consider only the really essential underlying facts of structure, and that all such structural features as were merely dependent upon adaptation to some particular mode of life should be disregarded. Thus, Linnæus had grouped the whales and dolphins (Cetacea) with the fishes, the ground of this arrangement being that both the whales and the fishes live in a watery medium, and are there-

* Owen, 'Lectures on Invertebrate Animals,' p. 9.

fore similar to one another in the general form of the body and in certain other external features.

On the other hand, Cuvier showed that, as regards their whole organisation, and more especially as regards their anatomical structure, the whales and dolphins are related, not to the fishes, but to the ordinary quadrupeds, and that therefore they should be regarded as a group of Mammals specially modified for an aquatic life. Acting upon this principle, we may say with Owen that 'the characters of the classes of animals have been rendered by the immortal Cuvier, the highest expression of the facts ascertained in the animal organisation.' It is not meant by this, of course, to assert that Cuvier's classification was by any means perfect, for we have seen that it was not; nor that he was always correct in his views as to the facts concerning the anatomy of animals; nor even that he always applied his own principles strictly. An absolutely perfect classification will only be possible when we are acquainted with all the facts as to the organisation of every animal—in other words, it will never be possible. All that can be asked of a classification is, that it should be the formal expression of the known facts of comparative anatomy at the time when it was drawn up. Cuvier recognised the true principles of all philosophical classification, and being admittedly the first comparative anatomist of his day, he was able to construct a classification of the animal kingdom which was an immense advance upon anything that had preceded it, and the main outlines of which still endure. Since Cuvier's time, however, comparative anatomy has made incalculable strides, and it is therefore no disparagement to Cuvier's pre-eminent merits, to say that naturalists are at present able to construct a system of classification which in many respects is greatly in advance of the arrangement proposed by the great French zoologist. Moreover, there is one department of zoology which in Cuvier's time was almost nonexistent, and which has the most important bearings upon the classification of animals—namely, the department of Embryology. It is now possible to supplement the knowledge gained from an anatomical examination of the bodies of adult animals by an investigation into the various changes, anatomical and physiological, which precede the attainment of the adult condition. The true relationships of animals thus become much more clearly recognisable than they can be when we have to compare together only the much modified and specialised structures of the fully-developed organisms.

Apart altogether from the merits of his system, or from his recognition of the principle that comparative anatomy is the true basis of scientific classification, Cuvier's contributions to morphology are of the most extensive kind. His 'Leçons d'Anatomie comparée' was the first systematic treatise upon the science of comparative anatomy; for John Hunter had not been able to publish any complete work on this subject. In certain special departments, Cuvier's anatomical researches form the basis of everything which has been since accomplished. We may instance more particularly his contributions to the comparative anatomy of the Molluscs, the osteology of the Mammalia, and the fishes. As regards the last of these groups, Dr Günther, one of the most eminent of living ichthyologists, has remarked that 'the investigation of their anatomy, and especially of their skeleton, was taken up by Cuvier at an early period, and continued till he had succeeded in completing so perfect a framework of the system of the whole class, that his immediate successors could content themselves with filling up those details for which their master had no leisure.

Lastly, Cuvier effected the most important advances as regards the natural history of former periods of the development of the earth. Indeed, it is not too much to say that Cuvier may fairly claim to have been the chief founder of the modern science of 'Palæontology;' a science which has enormously expanded during the last fifty years, and which, when fully mature, will obtain the recognition to which it is justly entitled, of being from all points of view one of the most important branches of scientific zoology. It is true that palæontology is often spoken of—in some cases even by scientific writers—as a branch of geology. It has even been termed 'the handmaid of geology.' A more erroneous conception of the entire aim and scope of palæontology could not well be formed; and it is one which will certainly not be accepted by any one who is himself a palæontologist in the proper sense of the term. It may be freely admitted that many of the earlier palæontologists dealt with fossils to a large extent from the geological point of view, rather than in their zoological aspect. It is also true that some so-called palæontologists have been little more than 'collectors,' and have had no real grasp of the scientific side of palæontology. This is, however, equally true of zoology in its earlier days; and, even now, 'collectors' are by no means extinct, nor are their labours by any means to be despised. Recent zoology, also, has as one of its departments the 'Geographical Distribution of Animals;' but no one for this reason would think of asserting that zoology was only a branch of geography. The sole relations between the subjects of geology and palæontology arise from the fact that *fossils* occur in *rocks*. Geology is to palæontology almost precisely what geography is to zoology. In its essence, however, palæontology is concerned entirely with the study of the remains of animals and plants; and in its two divisions of palæozoology and palæobotany, it is a branch of zoology on the one hand, or of botany on the other hand. The very use of the separate term 'palæontology' is a misfortune and a cause of error. The truth is, that all that part of palæontology which is concerned with animals is a branch of zoology; and just as a man cannot be a good palæontologist unless he is first a good zoologist, so it may be safely stated that a man cannot be a good zoologist, unless he has at least a good general knowledge of palæontology.

Cuvier's palæontological researches were mostly carried out in connection with the numerous remains of animals, chiefly Vertebrates, which had been met with in the strata of the neighbourhood of Paris. With his usual love of thoroughness in all he did, Cuvier undertook, in conjunction with Alexander Brongniart, an investigation into the arrangement of the Tertiary strata round Paris, in which these fossil remains abounded, and the results of this investigation were published in 1808, in the famous joint memoir, entitled 'Essai sur la Géographie minéralogique des Environs de Paris.' This work may be looked upon as the most important contribution which had been made up to that time to the study of the Tertiary series of rocks. It was principally, however, by the study of the organic remains contained in these strata that Cuvier has made himself famous.

Long controversies had been carried out among the earlier naturalists as to whether 'fossils' were really the remains of animals at all—many holding that they were merely peculiar mineral structures, formed by a kind of 'plastic virtue' in the earth itself. The notion, however, that fossils were merely *lusus naturæ* had been given up before Cuvier's time by most of the leaders of science; though it was still generally held that fossils were the remains of the animals and plants now in existence upon the globe. It had, of course, often been pointed out—as early as the time of Hooke and Ray, in fact—that many fossil shells were quite unlike any similar shells now existing; but it had been common to meet this by the argument that our knowledge of living animals was still very imperfect, and that very probably further investigations would show that the fossil forms which were supposed to be extinct, were really still living in some hitherto unexplored region. This last argument was, however, rendered quite untenable by the discovery of the remains of numerous unknown quadrupeds, often of large size, in the Tertiary beds round Paris; since, even in the beginning of this century, it was certain that no noteworthy discovery of large living Mammals was likely to be made in any of the less known portions of the earth's surface.

Cuvier, however, went further than this. He showed, by a close comparative examination of the fossil bones of these Mammals with the bones of their nearest living allies, with which they had previously been confounded, that such differences existed between them as to render it certain that the fossil forms belonged to 'extinct' species. The one grand point in which Cuvier's views fell short of those of modern palæontology, was that he failed to recognise any direct connection, by modification or descent, between the extinct species of animals and those now alive. On the contrary, Cuvier, like all the geologists of his time, was a 'catastrophist.' In other words, he believed that the present period was separated from preceding periods—as these were supposed to be separated from each other—by sharp lines of demarcation, due to great 'catastrophes' or natural convulsions, by which the animals and plants of each period were destroyed, a new series of organic forms coming into existence at the commencement of each fresh period.

On the question, also, of the nature and origin of 'species,' Cuvier was entirely in agreement with most of the naturalists of his day, being a firm believer in their fixity and immutability. His great predecessor and contemporary, Lamarck, whose views will be more fully discussed hereafter, had attacked this complex problem from its zoological side; and had arrived at the conclusion that the existing species of animals and plants had been produced by the modification of pre-existing species. In this conclusion, however, Lamarck had run counter to the most cherished beliefs of zoologists generally; and the prejudices which he had to confront were not lessened by the fantastic theories which had been put forward upon this subject by some of his own countrymen.

Thus, Benedict de Maillet, who wrote theoretically, and without any special knowledge of zoological science, had published a curious work entitled 'Telliamed,* or a Discourse between an Indian Philosopher and a French Missionary on the Diminution of the Sea,' of which an English translation was published in 1749. The fundamental proposition of this work was that the sea had at one time covered the whole of the dry land, and that therefore all the primitive forms of animal life must have been marine and aquatic in their habits. Hence, he supposed that the inhabitants of this hypothetical universal ocean had become changed into new forms, when the sea had retired, and the land had come into existence. In this way, all our present diversified forms of animal life had been produced, some of those animals which lived near the surface of the sea (such as Flying-fishes) becoming developed into birds; while some of those which lived at the bottom of the sea became converted into the terrestrial quadrupeds.

Again, Robinet, another theorist, had likewise published, in 1768, a work entitled 'Considerations philosophiques sur la Gradation naturelle des Formes de l'être,' in which he endeavoured to establish the proposition that the lower animals were merely the unsuccessful attempts which Nature had made in the production of Man.

In his views as to the fixity of species, Cuvier, on the other hand, was strictly orthodox. He recognised the existence of numerous *varieties*, especially among the

* Telliamed is an anagram of the author's own name.

domesticated animals; but he regarded the peculiarities of these as purely superficial. He believed that these varietal differences were the result solely of differences in the external conditions to which different individuals of the same species were exposed. He regarded these differences, therefore, as evanescent, and he would certainly have rejected entirely the idea that 'varieties' are 'incipient species.'

In dealing with the views which had been put forth by Lamarck, it seemed to Cuvier a sufficient argument that we find in the Catacombs of Egypt the mummies of cats, dogs, monkeys, and other animals, in a state of excellent preservation; and that we can therefore be certain that these particular species have remained essentially unchanged during the long period, from a human point of view, which separates the formation of the Catacombs from the present day. This argument, however, was sufficiently met by Lamarck himself. Lamarck saw no difficulty in accepting the fact that the species of animals preserved in the Egyptian catacombs are in all essential respects precisely similar to forms now in existence. On the contrary, 'it would assuredly be singular,' says he, 'if this was otherwise; since the position and climate of Egypt remain at the present day almost precisely what they were in the epoch of the catacombs.' Hence the animals which now live in Egypt find themselves under exactly the same conditions as they were then, and have therefore retained the habitudes which they at that time possessed. Besides, he adds, there is nothing in the observation in question to prove that these animals have existed from the beginning in their present form. It merely means that they have remained unchanged for the last two or three thousand years; but every one who has been in the position to appreciate the antiquity of Nature, will readily give the proper value to such a period, as compared with the age of the world.

Apart from his views as to 'species,' it is to Cuvier that we owe the establishment of a really scientific basis to palæontology. Cuvier showed that the only method by which the remains of fossil animals could be scientifically investigated, was by comparing them morphologically with the known living forms. In other words, he applied to palæontology those principles of comparative anatomy which he had used with such brilliant results in his purely zoological investigations. There is, however, an obvious difficulty in the way of the application of the laws of comparative anatomy to fossils in the same way as they can be applied to animals now in existence. In the case of the latter, we have the entire organism before us; we have not only its skeleton, but also its muscles, nerves, blood-vessels, and internal organs generally. Moreover, in systematic zoology, it is from the soft parts, rather than from the skeleton, that we in many cases draw our most weighty conclusions. In fossil animals on the other hand, with very few exceptions, only the hard parts are preserved for our inspection and examination. If we are to draw any conclusions at all as to the relationships and systematic position of extinct animals, we must do so from the characters and structure of the skeleton alone. Besides, in the case of a great many fossil animals, it is not usual that even the entire skeleton is preserved. In the case of all the Vertebrate animals, at any rate, the palæontologist is usually called upon to frame his conclusions on a fragmentary specimen. He may have only a single bone or tooth; or he may have a number of detached bones. Only rarely does he find a complete skeleton, or meet with the bones still in their proper places and connections.

This inherent difficulty in all palæontological investigation was solved by the establishment by Cuvier of the famous law of 'the correlation of organs.' Cuvier showed that certain organs or structures in animals are only found in association with one another; so that if one of these correlated organs be found to be present, then we may be sure that the others will also be there. In some cases, this correlation or association of particular organs is based upon an obvious physiological connection. Thus, thin-walled hollow bones are associated with a peculiar form of lung, in which the greater air-passages (bronchi) do not end within the lung itself, but become connected at the surface of the lung with membranous receptacles or air-sacs distributed in different parts of the body. Again, the peculiar form of toe-bone which is adapted for the carrying of a hoof is correlated with such other modifications of the bones of the limb as are needed to secure the absence of rotation in the bones, and to insure the fitness of the leg for its special function of supporting the weight of the body. In very many cases, however, no physiological explanation can be given as to the association or correlation of particular organs. Thus, all animals in which the skull is jointed to the backbone by a double articulation, and in which the two halves of the lower jaw are composed each of a single piece, have at the same time the glands by which they are enabled to suckle their young. All animals possessing these two structures also possess (or may possess) the special integumentary appendages known as hairs. Similarly, all those animals which have the stomach adapted for chewing the cud, or ruminating, have as a correlation with this, no more than two functionally useful toes, the third and fourth toes. All such animals, moreover, have an incomplete development of the incisor teeth in the upper jaw. They are also the only living quadrupeds which have horns developed upon the frontal bones.

The few examples given above may suffice to illustrate the general nature of the 'law of the correlation of organs.' Stated in its most general form, this law asserts that all the parts of the organism stand in some relation to each other, the form and characters of each being in direct connection with the form and character of all the rest. The nature of this connection is in many cases hidden from us; but it is certain that if, by an arbitrary exercise of will, we could suddenly change the form of any one organ in any given animal, we should find ourselves compelled to make changes in all the other organs of the same animal. In many cases, perhaps, the changes necessitated by the modification of some particular organ might be slight; in most cases we should be unable to see why any changes should be needed at all, beyond the one with which we had started; in all cases the fact would remain, that the living organism is an aggregate of parts so put together that any modification of any one part necessitates a modification of all the rest.

The application of this law to palæontology is easy to

understand. It is found from the study of living animals, where all the parts of the organism can be investigated as a whole, that certain organs or structures are found associated with one another, or, at any rate, are never found apart. In the case of fossil animals, we never have more than certain parts of the organism preserved. We never, save in such exceptional instances as the preservation of the bodies of animals in the frozen soil of Siberia, have the opportunity of examining the *whole* organism. By the help, however, of the law of 'the correlation of organs,' we can 'reconstruct' the animal from its fragments. If we find certain structures preserved in the fossil, we can infer that certain other correlated parts must have been present. Thus, from a single molar tooth it may be possible to infer the form of the jaw, the structure of the limbs, and, in fact, the general features of the organisation. Those who wish to learn with what precision and certainty an extinct animal may in this way be 'reconstructed' from its fragmentary remains, can easily satisfy their curiosity by reference to the pages of the 'Ossemens fossiles' of Cuvier, or to the works of his illustrious disciple, Sir Richard Owen.

It should be pointed out, however, as indeed was recognised by both Cuvier and Owen, that the law of correlation of organs can only be applied in practice with certain reservations, of which the following are the most important: In the first place, the law is a purely empirical one, and is based wholly upon the results of observation and experience. Having, therefore, no rational basis, it is always liable to be overthrown in particular instances by more extensive observation, though its validity as a general law remains unaffected. Thus, so far as our present knowledge goes, milk-glands are never found except in association with a double joint between the skull and the backbone, and a simple structure of the lower jaw. But it is quite *possible* that we might some day find mammary glands in association with a differently constructed jaw, or with a single occipital articulation, however unlikely this may seem at present. Even living animals are constantly demonstrating to us the danger of these empirical generalisations. Thus, until recently, it might have been safely asserted that all Vertebrate animals with a bony skeleton, which produced their young as eggs, had also a lower jaw in which each half was composed of more than one piece. We know now, however, by recent investigations that the duck-mole and spiny ant-eater, both of which have the lower jaw simple, are oviparous animals. Extinct animals still more forcibly exemplify the necessity for caution in reasoning from the presence of one structure that another correlated structure was present. Thus, until a few years ago, it would have been unhesitatingly admitted that the possession of a covering of feathers was correlated with the possession of a horny covering to the margins of the jaws, and therefore with the absence of teeth. It would also have been admitted that feathers were correlated with saddle-shaped faces to the bodies of the vertebræ of the neck. Through the researches of Professor Marsh, we are, however, now acquainted with extinct birds which must, as birds, have possessed feathers, but in which the jaws were furnished with teeth in sockets (the *Odontornithes*). The same distinguished observer has also brought to light an extinct bird (*Ichthyornis*), in

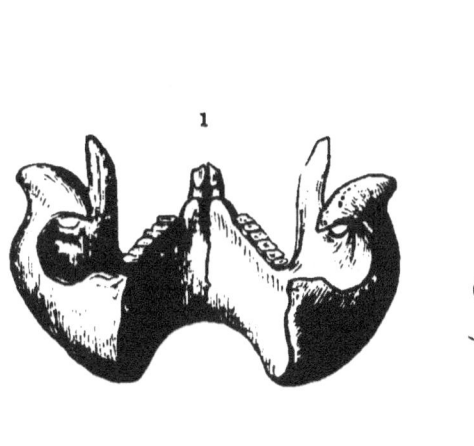
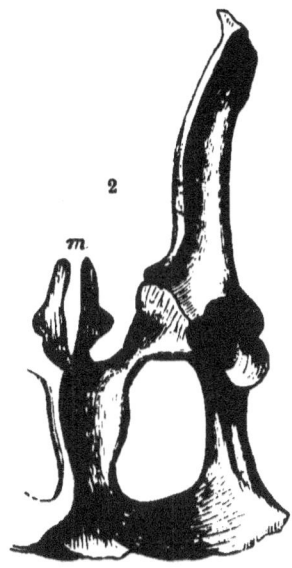

1. Lower jaw of Wombat, showing the 'inflected' angle of the jaw.
2. Pelvis of a Kangaroo, showing the 'marsupial bones' (m).

which not only were teeth present, but the faces of the neck-vertebrae were cupped, as they are in fishes, instead of being saddle-shaped as they are in normal birds. Again, in all living animals in which hollow bones filled with air are present, the skin is furnished with feathers; but the extinct Pterodactyles, or flying reptiles, possessed hollow bones, while there is no reason to think that their skin was feathered.

In the second place, when we have to deal with fossil organisms, we may easily assume that a particular structure was absent in an animal, whereas it might have been present, and yet might not have been preserved, owing to its only having been present in a condition incapable of preservation in the fossil condition. Thus Marsupial quadrupeds (kangaroos, opossums, &c.) may be stated, as a general rule, to have the lower jaw of a characteristic form, the part of the jaw known as the 'angle' being bent inwards or 'inflected.' Correlated with this peculiar structure of the jaw, but having no recognisable connection with it, are two little bony splints, which are attached to the brim of the pelvis, and which are known as the 'marsupial bones.' No exceptions were known in the time of Cuvier to this rule; hence Cuvier was entitled to regard this as a constant correlation. Thus he met with a fossil skeleton of a quadruped, like all such fossils, only preserved in parts, which showed the lower jaw; and finding that the angle of the jaw was 'inflected,' he came to the conclusion that it was a Marsupial. Moreover, from the structure of its teeth he inferred that the skeleton belonged to one of the opossums, such as now inhabit the American continent, and he named it the *Didelphys gypsorum*.

As all living Marsupials are found in Australia, New Guinea, certain of the islands of the Pacific, and in North and South America, the alleged discovery of an opossum in the Tertiary strata near Paris, naturally excited some incredulity in the scientific world. In order to dissipate this incredulity, Cuvier invited his scientific colleagues to meet him, and proceeded in their presence to cut away with a chisel the stone enveloping the bones, so as to bring into view the front part of the pelvis, which lay deeply buried in the matrix. On accomplishing this, he was able to demonstrate at once that the pelvic bones carried the 'marsupial bones,' which are so characteristic of the opossums and of Marsupial quadrupeds in general.

The Fossil Opossum of Montmartre (*Didelphys gypsorum*), showing the 'marsupial bones' attached to the brim of the pelvis (*m*).

It is needless to add that this demonstration of the value of the law of the correlation of organs in palæontological researches excited the greatest admiration, and was regarded as absolutely conclusive. In one important point it *was* indeed conclusive; since no known animal outside of the order of the Marsupials is known to have *both* an inflected angle to the lower jaw, and also marsupial bones on the brim of the pelvis.

The conclusiveness arose, however, from the fact that Cuvier found *both* these structures together; and we now know that the presence of the one would not necessarily prove the presence of the other. Cuvier knew this himself so far as the presence of 'marsupial bones' is concerned, because he knew that these bones occur in the duck-mole and spiny ant-eater, in which the angle of the jaw is nevertheless not inflected. He did not know that he might have found the inflected angle of the lower jaw, and that he might have been quite right in his conclusion that the animal was a Marsupial; and yet, on laying bare the pelvis, he might have found no 'marsupial bones.' It is known, namely, that in certain living Marsupials (the *Thylacinus* of Tasmania) the 'marsupial bones' do not become converted into bone, but remain permanently in the condition of cartilage. These structures would therefore be absent in any fossil specimen of such a Marsupial, since cartilages are not preserved in the fossil state. Hence, it is possible, though not probable, that we might some day meet with the skeleton of some extinct Marsupial, in which we should find the angle of the lower jaw to be inflected, but which would nevertheless show no traces of 'marsupial bones.'

In the third place, in any two correlated organs it is not usual that each is correlated with the other, but that one of the two is correlated with the other. That is to say, of any two correlated organs, A and B, it may be true that A is never found without B, but it does not follow that B may not occur without A. Thus, the presence of a stomach adapted for 'rumination' is invariably associated with an imperfect development of the incisors of the upper jaw, the central upper incisors being always wanting; but it is not the case that an incomplete condition of the upper incisors, or the absence of the central ones, is necessarily correlated with the habit of chewing the cud. The proper way of putting the case is to assert that certain structures (A) are never found apart from other structures (B), though the latter may be present without the former. When, therefore, we find a lower jaw having its angle 'inflected,' we may, with our present knowledge, assert that the animal to which that jaw belonged must have possessed 'marsupial bones' or 'marsupial cartilages' upon the brim of the pelvis. If, however, we were to find a pelvis with marsupial bones, we should not be justified in asserting that the owner of the same must have possessed an inflected angle to the lower jaw. On the contrary, we know that such an assertion would be erroneous, since the 'marsupial bones' are present in the duck-mole and spiny ant-eater, in which the angle of the jaw has its usual form.

RETROGRESSION.

SWAINSON AND THE CIRCULAR CLASSIFICATION.

WE have now considered the main features of the work effected by Cuvier in zoology, and we have next to take a glance at the retrograde and in all respects singular system known as the Circular Classification. The original author of this system was William Sharpe Macleay, a well-known and able entomologist, and the first complete exposition of his system is to be found in a very rare work entitled 'Horæ Entomologicæ,' published in London in the year 1819. Though Macleay was the inventor—the term may be used advisedly without disparagement of his undoubted talents and perfect good faith—of the 'Circular System' of classification, Swainson, the well-known ornithologist, was its principal expositor; and it is therefore most suitable that we should consider this system in connection with the latter rather than the former naturalist. In the first place, then, let us take a glance at Swainson's life as told by himself.*

William Swainson was born in October 1789, his father

---

* 'Cabinet Cyclopædia: Biography of Zoologists,' pp. 338–352.

being an official in the Custom-house. His ancestors had been for generations 'statesmen' in Westmorland, and had lived on their property near Hawkshead; but the family estate had gradually passed into other hands. Swainson was brought up by his father with a view of entering the Custom-house, and his early education was cut short, in consequence of his having an impediment in his speech which entirely prevented him from studying languages, and also because he does not appear to have had 'the least aptitude for the ordinary acquirements of schools.' Hence, we find him at the age of fourteen as junior clerk in a secretary's office in the Customs, with a salary of £80 a year. He had, however, a passion for natural history, a rooted dislike of official trammels, and a burning desire to travel. After he had been about three years a clerk in the Customs, his father obtained for him, in 1807, an appointment in the commissariat department of the army, and in the spring of the same year he was despatched to join the Mediterranean army in Sicily. Here he remained several years, and as his duties were light, he was able to wander all over Sicily, collecting animals and plants, and also to visit Greece. Subsequently he was quartered in Italy, first in one city, and then in another; but his health became impaired, and in 1815 he was sent home on sick leave.

Swainson had now risen to the rank of assistant commissary-general on the staff of the Mediterranean army, and as he was only twenty-six years old, he might have expected much higher promotion had he remained in the service. He had, however, never taken kindly, as some men never do, to official life. He could not even endure the restraints of English society. 'I had,' says he, 'to join dinner-parties, drink wines which I detested, ride in carriages, dance at balls, and do a hundred other things for which I had neither health nor inclination.' His old passion for foreign travel returned irresistibly upon him. He gave up his appointment in the army, and went upon half-pay. At first he thought of visiting Cape Colony; but hearing that the traveller Burchell had just returned thence with a collection of objects of natural history which filled two wagons, he rashly concluded that there would be nothing left for him to collect there, and he abandoned his intention. It happened however, that at this time Mr Koster, who had formerly travelled in Brazil, and had published an account of his travels, was about to return to that country; and Swainson forthwith made up his mind to accompany him. After some time spent in South America, travelling about and making zoological collections, Swainson returned to England, and settled down to study his collections and describe the results. He was shortly thereafter elected a Fellow of the Royal Society, though he does not seem to have met otherwise with much encouragement. With characteristic energy, he determined to learn the then newly introduced art of lithography, and to see how far this process could be utilised in the production of plates of animals suitable for colouring. His attempts in this direction proving successful, he issued a series of descriptions and figures of new, rare, or remarkable animals, under the title of 'Zoological Illustrations.'

Swainson now settled in London, where he worked hard for two or three years, and followed up his former attempt at authorship, by the publication of some numbers of illustrations and descriptions of exotic shells. An attempt to obtain an appointment in the zoological department of the British Museum was not successful, and Swainson, having now married, found it necessary to increase his income. He determined, therefore, to become a professional author, and was fortunate enough to form a connection with the great publishing house of Longman, Orme, Brown, & Co. The next few years of Swainson's life were occupied in incessant literary and scientific activity. The principal works which he gave to the world are the 'Cabinet Cyclopædia of Natural History,' the portion of the 'Fauna Boreali-Americana' dealing with the birds, a second series of the 'Zoological Illustrations,' and three of the ornithological volumes in the 'Naturalists' Library.'

In 1835, Swainson lost his wife, and he subsequently determined to emigrate with his family to New Zealand. This determination he carried into effect, and with this closed his scientific career. He died in New Zealand in the year 1855.

No doubt can be entertained as to Swainson's having possessed considerable natural abilities. Had he received a rigorous and methodical training in early life, he would probably have left a much more abiding mark upon zoological science, and have occupied a more conspicuous place in the long list of British naturalists. As it is, he attained a high reputation in certain departments, and especially in ornithology. It is, however, unnecessary here to discuss the value of Swainson's observations in this last, or in any other of the many branches of natural history upon which he wrote so copiously. He is at the present day best known, perhaps, in consequence of his close connection with the 'Circular System' of classification, with which he had entirely identified himself. This theory we may therefore now proceed to discuss very briefly; and in so doing, it will be best to consider Swainson's enlarged and amended scheme, rather than the comparatively rough outline of the circular classification given by Macleay, the actual author of the system, and the originator of the notion of circular affinities.

In expounding his classification of the animal kingdom, Swainson, to begin with, discusses fully what he conceives should be the principles upon which a natural arrangement of animals may be founded. He points out that the likenesses which subsist between different animals are not only different in degree, but also different in *kind;* and he divides such likenesses into what he terms likenesses of 'analogy' and likenesses of 'affinity.' What he understands as likenesses of 'analogy' are all such likenesses or resemblances between different animals as are either what we may call accidental, or which depend merely upon similarity of the mode of life. Thus, he instances the likeness between the striped skin of the tiger and that of the zebra as an example of analogy. Similarly, the likeness between the whales and the fishes, dependent upon the purely aquatic life of both, he instances as a likeness of analogy. On the other hand, Swainson recognised that there exist other likenesses, affecting the whole organisation of the animals compared, of more real and fundamental character, and these he termed likenesses of 'affinity.' As an example of these, he instances the likenesses between the tiger and the cat, which affect all the more important features of the anatomical structure of these two animals.

Though Swainson recognised in theory the distinction between these two kinds of likenesses—now known respectively as likenesses of 'analogy' and likenesses of 'homology'—he showed, by his loose application of these in practice, that he did not sufficiently recognise the causes of the distinction between them. As modern naturalists understand this matter, likenesses of 'analogy' are purely *physiological* or adaptive, and depend merely upon similarity of mode of life or external environment. Thus, whales are like fishes because both have certain structural modifications, as regards the form of the body and shape of the limbs, which adapt them for a life in a watery medium. Similarly, there is a certain resemblance between bats and birds, due to the fact that both are adapted for flight in the air. These physiological or analogical likenesses, however, are quite independent of real relationships, and they are therefore of no value for classificatory purposes. On the other hand, there are likenesses between animals which are *morphological*, and which are quite independent of the kind of life which the animal may lead, or the nature of its surroundings. These 'homological' likenesses are dependent upon identity of structure and fundamental plan, and they exist irrespective of, and despite of, the animal's habit of life or the particular use to which it may put its organs. Thus, to give a single example, there exists a homological likeness, due to identity in fundamental plan of construction, between a butterfly and a lobster; though these animals are by no means like one another in external appearance, and are adapted to entirely different modes of life. It is in the separation of the merely physiological or adaptive characters of an animal from its really essential morphological characters that a great part of the work of the scientific zoologist consists; and it is also upon characters of the latter class that all modern systems of classification of the animal kingdom are based.

To return, however, to Swainson and to the circular classification. To some extent, Swainson undoubtedly recognised the underlying distinction between these two kinds of likenesses among animals. He also recognised that all classifications which are based upon likenesses of analogy are necessarily 'artificial,' and that the basis of a 'natural' classification can only be found in the 'affinities' or homological likenesses between animals, since these alone are indicative of true relationships. It is singular that, starting with comparatively clear ideas as to what points were of really taxonomic value, Swainson should have given his adhesion to one of the most fantastic and unnatural systems of zoological arrangement which have ever been promulgated. For our present purpose, it is enough to give a mere outline of this system, which Swainson laid down in the following propositions:

(1) The entire series of animals is a continuous one, forming a circle; so that, 'commencing at any one given point, and thence tracing all the modifications of structure, we shall be imperceptibly led, after passing through numerous forms, again to the point from which we started.' In accordance with this proposition, Swainson divided the entire animal kingdom into five great groups,

which he arranged in a circle, as shown graphically below.

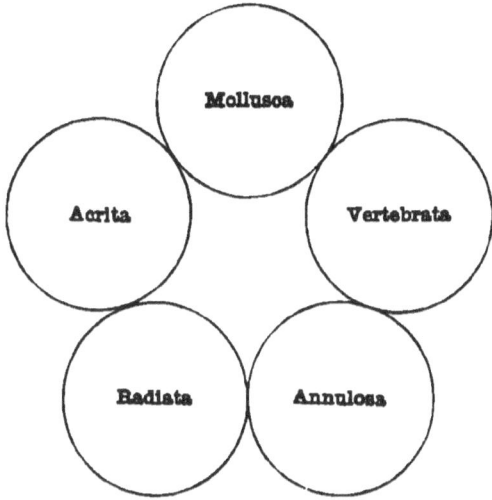

In precisely the same way, Swainson divided the series of the Vertebrate animals into a chain of circular groups, as follows :

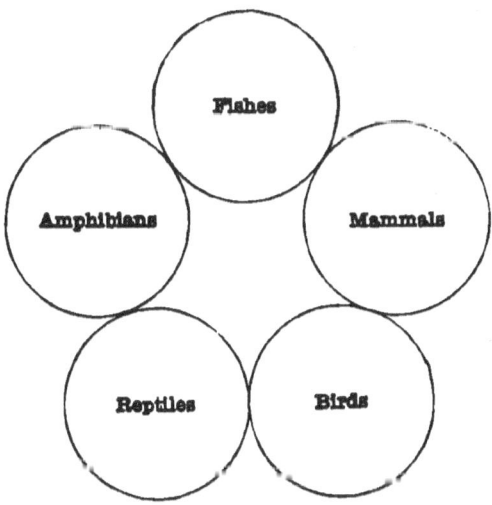

It would be a waste of time and space to point out in detail the extraordinary inconsistencies of such a circular arrangement as exhibited in either of the above schemes. It is enough to point out that passage along such a circular series as either of the above is easy enough and natural enough in a certain direction, and up to a given point; but that it is an impossibility to *complete* the circle, in consequence of an absolutely impassable gap between two of the groups in the series. Thus, in the circular arrangement of the whole animal kingdom, as given above, one may pass, without any violation of natural affinities, from the Acrita (the Protozoa of modern naturalists) through the sponges to the Radiata (the Cœlenterate animals and Echinoderms). Through the Echinoderms one may pass from the sea-cucumbers (Holothurians) to the spoon-worms, and thus into the Annulose series; and the passage from the Annulose animals to the Vertebrate animals is at any rate a conceivable one. Similarly, in the backward progress, one may pass naturally enough from the Vertebrates to the Molluscs, through the cuttle-fishes (Cephalopods). In order to *complete* the circle, however, there remains the final step of passing from the Molluscs to the Protozoa (the Acrita), two groups separated by a hiatus unbridged by any intermediate form.

Similarly, in the circular series of the Vertebrate animals, starting with the Amphibians (frogs and newts) as the assumed lowest group, we pass naturally enough to the reptiles, and from the reptiles to the nearly related group of the birds. From the birds one may get to the Mammals by the help of the oviparous duck-mole and spiny ant-eater. Or, if we pass in the other direction, we travel quite naturally from the Amphibians to the fishes. In either case, however, in order to complete the circle, one has to get from fishes to Mammals, or *vice versâ*. The gap thus caused is, however, a hopeless one, since the really intermediate groups of the reptiles and the birds have been left *on the other side of the circle*.

(2) The second proposition of the circular system is, that 'the primary circular divisions of every group are three actually, or five apparently.' Differences of opinion have arisen among the advocates of the circular classification as to the number of groups in each division, though all were agreed that the system was based upon *some* fixed number, which governed all the subdivisions from the highest to the lowest. Macleay, the founder of the system, thought there were five main divisions and five smaller ones (making ten in all). Fries, the botanist, adopted four as the number governing his classification of plants; but as he admitted that his central group could always be split into two, this practically made his system likewise quinary. Swainson, as seen in the above-quoted proposition, adopted three as his number; but as one of these groups was supposed to be always divisible into three smaller sections, this also was practically identical with taking five as the ruling number. Other 'circularists,' again, adopted seven as the governing number. Five was, however, the number most usually adopted, and for this reason this system has often been spoken of as the 'Quinary Classification.' According to this, therefore, everything must go by fives. If animals obstinately refused to range themselves into

L

fives, this could only be because we knew too little about them to make them do so. Thus, as shown above, there were five sub-kingdoms, or primary divisions, of the animal kingdom. Each sub-kingdom was similarly divisible into five classes, as shown above in the case of the Vertebrates (in which there really are five classes). Each class fell to be divided into five orders, each order into five families, and so on.

As just mentioned, however, Swainson did not accept this quinary arrangement without some modification. He thought that *three* was the governing number; but then he supposed that one group could always be divided into three smaller circles. In any three groups—forming a closed system—one group is what he called 'typical,' a second is 'sub-typical,' and the third is 'aberrant'; but the aberrant group itself forms a closed system of three smaller circles, thus:

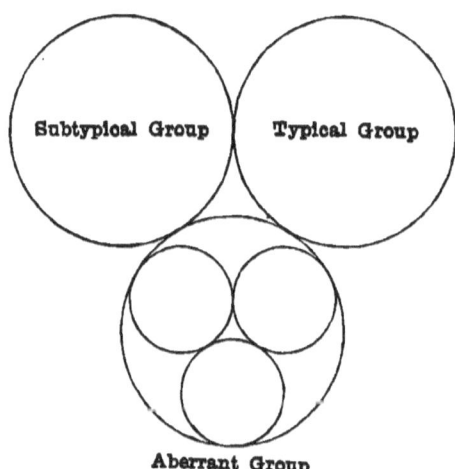

We may illustrate this neat mathematical arrangement of animals by the Vertebrates:

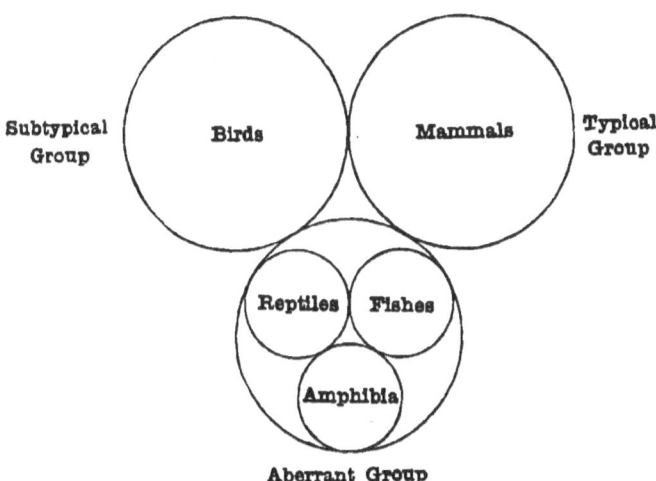

(3) The third, and perhaps the most fundamental, proposition of the circular system is, the animals contained in any given circular group are 'symbolically or analogically represented' by the animals contained in each and every other circular group in the animal kingdom. In order to understand this proposition fully, a few words must first be said on what Swainson understood by his 'typical,' 'sub-typical,' and 'aberrant' groups. In any given series of animals the 'typical' group is that comprising those forms which possess the largest number of the distinctive characters peculiar or common to the whole series. The 'sub-typical' group comprises those forms which exhibit a smaller proportion of the presumed distinctive characters of the series. Finally, the 'aberrant' group comprises forms which

have the fewest of the characters common to the series, and which therefore necessarily possess many characters common to *other* groups of animals. Moreover, Swainson considered that every 'aberrant' group exhibited three principal modifications of structure, which constituted the basis of as many minor groups. One of these subordinate sections of the aberrant group was supposed to contain animals adapted for an aquatic life. The animals of a second section were supposed to be adapted for obtaining their food by suction; and a third or 'rasorial' section was formed for the reception of types adapted for scratching or for climbing. Thus, as shown in a preceding diagram, the Mammals are the 'typical' group of Vertebrates, and the birds are the 'sub-typical' group; while the 'aberrant' group comprises the 'aquatic' section of the fishes, the 'suctorial' group of the Amphibians, and the 'rasorial' group of the reptiles.

Now starting with this basis—admitting, that is, that you could divide a given series of animals into three primary groups, a typical, a sub-typical, and an aberrant group, and admitting further that the last contains three minor groups, one aquatic, one suctorial, and one rasorial—the fundamental proposition of the circular classification is, that *every* series of animals can be similarly divided, and that each quinary system, however small, must *represent* each and every other system. If it does not do so, it is, *ex hypothesi*, not a natural group. By way of exemplifying this proposition, we may present here the tabular view which Swainson gives of what he regards as the 'beautifully simple and harmonious' analogies between the circular system of the birds and that of the Mammals:

Primary Types.	Orders of Birds.	Typical Characters.	Orders of Mammals.
1. Typical........	*Insessores* (Perchers).	Organs of prehension and general structure highly developed.	*Quadrumana* (Monkeys).
2. Sub-typical...	*Raptores* (Birds of Prey).	Carnivorous, claws retractile.	*Feræ* (Beasts of Prey).
3. Aberrant (aquatic group).	*Natatores* (Swimmers).	Live and feed in the water. Feet short, or none.	*Cetacea* (Whales, Dolphins, &c.)
4. Aberrant (suctorial group).	*Grallatores* (Waders).	Jaws much prolonged; burrow for their food.	*Glires* (Rodents and most Marsupials).
5. Aberrant (rasorial group).	*Rasores* (Scratchers).	Head with crests of horn or feathers; habits domestic; feet long, formed for walking.	*Ungulata* (Hoofed Quadrupeds).

Swainson excuses himself for not entering into long details explanatory of the above table, on the ground that explanation is unnecessary, seeing that 'the analogies are so perfect, and the series so completely in unison with those of all other animals.' To the modern scientific student it seems equally unnecessary to discuss such a table, as it violates all those fundamental canons of classification which have been established by the combined labours of naturalists for the last two centuries. It would not, indeed, be easy to produce a classificatory table of Mammals and birds more entirely at variance with what naturalists at the present day believe as to the true relationships of animals. One fact sufficiently proves this—namely, that in deference to the mystical 'quinary' law, the Mammals are divided into only five groups or orders; whereas naturalists consider that there are at least fifteen or sixteen natural orders of these animals.

The 'circular classification' is, then, a mere figment of the human mind; the notion of a quinary, ternary, or septenary division of animals is equally a product of the imagination. So far as our present knowledge goes, two things are abundantly evident. One of these is, that no numerically symmetrical arrangement of animals can, by any possibility, accord with their natural affinities and relationships. The other is, that any pictorial representation of the different groups of the animal series in the order of their natural alliances would assuredly not present us with a system of similar closed circles, but with a branched and ramified genealogical tree. One main trunk we should undoubtedly find; and this would give off numerous lateral stems, which would in turn subdivide, some branches ascending in the course of their development, while others, in consequence of degeneration, would descend. No numerical law could, however, possibly be formulated which would express the branching of the main stem of life; nor would there be any parity of size, or equality in zoological value, between the different branches of the parent trunk.

BRITISH ZOOLOGISTS
(CONTINUED).

In the twenty or thirty years which followed the publication, in 1817, of the 'Règne Animal,' the study of natural history was prosecuted in Britain by many distinguished and able men, and great advances were effected in almost all branches of the science. These advances, however, mostly concern our knowledge of special groups of animals, and are therefore of little interest except to specialists. As regards philosophical zoology, or the principles of natural history, the condition of the science remained without any noteworthy alteration, very much as it had been left by Cuvier. All that can be here attempted, therefore, is to give a very brief general sketch of the progress which was made during this period as regards special groups of the animal kingdom, with more particular reference to any point of exceptional interest. Many well-known names are, of course, necessarily omitted in such an outline, and any attempt to analyse the varied works which appeared at this epoch would lead us into paths which are only pleasant to walk upon for the initiated. The *typical* naturalist of this period is Edward Forbes, who may be selected for a longer notice.

As regards 'general zoology,' this period produced a number of able workers. No one, perhaps, contributed more largely to the diffusion of a knowledge of, and a taste for, natural history than the well-known Dumfriesshire baronet, Sir William Jardine, who united to a wide general knowledge of natural history and of geology a special acquaintance with ornithology. Sir William is probably best known as the editor of that excellent and popular work the 'Naturalists' Library,' of which more than one edition was published. This work consists of forty volumes, dealing with Mammals, birds, fishes, and insects, and illustrated with spirited figures, drawn and engraved by Lizars. Sir William Jardine was also joint-editor, along with Mr Selby and Dr Johnston, of the 'Magazine of Zoology and Botany,' of which only two volumes appeared (in 1837 and 1838); when it became merged in one of the most admirable of our existing zoological periodicals—namely, the 'Annals and Magazine of Natural History,' affectionately known to its readers by the abbreviated name of the 'Annals and Mag.'

The only general work treating of the whole of the animals of the British area, which appeared during this period, was the 'History of British Animals,' by the Rev. John Fleming, D.D., who was one time minister of Flisk in Fifeshire, and subsequently professor of natural philosophy in King's College, Aberdeen (from 1834 to 1845). This work was published in 1828, in one volume, and dealt systematically with all known British animals— a gigantic task for one man, even at that time, and an almost impossible one for a single worker now. It is unfortunately not illustrated, and the account given of the lower Invertebrates is necessarily meagre and imperfect. Fleming also wrote a work on 'The Philosophy of Zoology,' in two volumes, and a small treatise on the *Mollusca*, which appeared first in the seventh edition of the 'Encyclopædia Britannica,' and was published later (1837) in a separate form.

As regards the special group of the quadrupeds, perhaps the most important work published in England at this time was the 'Fauna Boreali-Americana,' or 'Northern Zoology,' of Sir John Richardson, which appeared in 1828. This noble work treated of the zoology of the northern parts of British North America, and was based upon the collections which had been gathered together during the northern land expeditions under the command of Sir John Franklin. The four quarto volumes are admirably illustrated, and treat respectively of the Mammals, the birds, the insects, and the fishes; the insects being described by Mr Kirby, while Swainson took part in the production of the volume on the birds. Unlike the 'Arctic Zoology' of Pennant, the 'Fauna Boreali-Americana' was the work of a naturalist who had personally visited the region of which he described the animals. Sir John Richardson had, in fact, been the chief surgeon and naturalist to the overland Arctic expedition of Sir John Franklin in 1825. In 1848, he once more visited the Arctic regions, in the hope of discovering, if possible, what had been the fate of the last disastrous expedition of Franklin. Richardson was a Scotchman, born at Dumfries in 1787. His life was one of great interest, but cannot be more particularly noticed here. He died in 1865.

The only other work dealing with Mammals, that may be noticed as treating exclusively of British types, is the 'History of British Quadrupeds,' by Professor Thomas Bell, at one time professor of zoology in King's College, London. This excellent work appeared in 1836, and is one of the admirable series of illustrated works on British zoology issued, and still being issued, by the well-known scientific publisher, Van Voorst. To this same series, Professor Bell contributed two other equally excellent treatises—namely, the 'History of British Reptiles' (1829), and the 'History of British Stalk-eyed Crustacea' (1853).

Ornithology has ever been a favourite subject with naturalists, and the number of names which occur to any one who has occasion to look into the zoological literature of this period, as belonging to eminent ornithologists, is noteworthy. Yarrell, Macgillivray, Swainson, Eyton, Gould, Selby, Jardine, and Waterton are names which have the pleasant ring of familiarity to the ears of all naturalists, and in some cases to extra-zoological circles. Such works as Eyton's 'Monograph of the Anatidæ' (1838), or Selby's 'Illustrations of British Ornithology' (1821-34), are doubtless little known except to specialists; but every one who has dabbled in natural history is acquainted with Gould's magnificent monographs, if only on account of their inimitable illustrations. There must also be very few, even of those who are not zoologically inclined, who do not know something of Waterton, not through anything that he did in ornithology, for in truth he was not a scientific zoologist, but through that most delightful of books, the 'Wanderings in South America.'

The two most essentially British ornithologists of this period were William Yarrell and William Macgillivray. The former is the author of what may be regarded as the standard work on the birds of our country—namely, the 'Natural History of British Birds' (1839-43). Yarrell was also the author of the equally well-known 'Natural History of British Fishes,' the first edition of which was published in 1836. Macgillivray is perhaps best known as the author of the 'History of British Land and Water Birds,' now a scarce and expensive work, which was published in 1837, in five octavo volumes. He published various other treatises, of which the two best known are 'The Natural History of Deeside and Braemar,' published posthumously in 1855, and his 'Lives of Eminent Zoologists' (1834), which formed one of the volumes of the 'Edinburgh Cabinet Library.' Macgillivray occupied the chair of natural history in Marischal College, Aberdeen, from 1841 to 1853, and as an ornithologist has not perhaps received generally full justice.

Coming next to the Invertebrates, there are only three groups which may be noticed, namely the Molluscs, the Insects, and the Zoophytes; and even these can only be glanced at in the most cursory fashion. In the department of the Mollusca, we find many well-known authors such as Turton, Wood, Burrows, Broderip, &c.; but none of these call for special remark. The two names which are most familiar to conchologists and naturalists generally in connection with this period are those of Sowerby and Woodward—though the latter more properly belongs to a later time. The name of Sowerby, moreover, is that of a series of naturalists and artists, who devoted themselves especially to natural history, mineralogy, and botany; and of whom three were more particularly concerned with conchology. One of these is James Sowerby, originally an artist, who was born in 1757, and died in 1822. His great work is the 'Mineral Conchology of Great Britain,' which deals with the fossil shells of our islands. James de Carle Sowerby was the son of the preceding, and followed in his footsteps. He was born in 1787, and died in 1850, and continued the publication of the 'Mineral Conchology.' Lastly, in more recent times, conchologists have been indebted to George Brettingham Sowerby for a work on the 'Genera of Recent and Fossil Shells,' a 'Manual of Conchology,' and a 'Thesaurus Conchyliorum.'

There has, however, been no naturalist of the present century to whom conchological students in general have been more deeply indebted than to Samuel Woodward, who was born in 1821, and died in 1865, and who therefore does not strictly fall into the period now under consideration. His 'Manual of the Mollusca,' of which the first edition appeared in 1841, has a world-wide reputation as one of the most philosophical and comprehensive treatises on a single large group of animals ever published in such a moderate compass. It is a model of everything which such a manual should be.

Entomology—an even more favoured subject than ornithology—commanded many votaries during the period here in question; and among them some of the most distinguished entomologists Britain has yet produced.

Nothing is, however, here possible beyond the mention of some of the leading works which appeared at this time. One of the best known and most widely used of these is the 'Introduction to the Study of Entomology,' the first edition of which appeared in 1828, the authors being the Rev. William Kirby and Mr William Spence, both distinguished specialists in the department of entomology. Another famous work is the 'Introduction to the Modern Classification of Insects,' by Professor J. O. Westwood. This appeared in 1839, and being of a more technical character than Kirby and Spence's 'Introduction,' has become one of the standard works of the entomological specialist.

As regards purely British entomology, the most famous work of this time is Curtis's 'British Entomology' (1824-1840). This beautiful treatise, in sixteen octavo volumes, is illustrated by admirably drawn coloured figures of the insects and of the plants upon which they feed; and is still one of the standard works on the subject with which it deals. Another work of the same character, but only dealing with certain groups, is the 'Illustrations of British Entomology,' by James F. Stephens. Mention must also be made here of the admirable anatomical memoirs dealing with Insects and Myriapods (centipedes and their allies), by Newport, who likewise contributed the article 'Insecta' to Todd and Bowman's 'Cyclopædia of Anatomy and Physiology (1839).

Lastly, as regards the lower Invertebrate animals, and more particularly the Zoophytes (Cœlenterate animals), there are only two names belonging to this period which need special mention. One of these is that of Dr George

Johnston, a medical practitioner at Berwick-upon-Tweed, and a well-known naturalist. The work by which Dr Johnston is best known is his 'History of British Zoophytes,' which appeared in 1838, and of which a second edition was published in 1847. As the Cœlenterate animals were, at the time when Johnston wrote his treatise, but imperfectly separated from other animals, we find here descriptions and figures not only of the British species of the Zoophytes strictly so called (namely, the Sea-anemones, the Sea-firs, &c.), but also of the Sea-mats and their allies (the Polyzoa). Moreover, it had not at this time been discovered that there was any connection between the jellyfishes and the ordinary plant-like Zoophytes, and the former found, therefore, no place in Johnston's work. The 'History of British Zoophytes' will always have to be consulted by any British naturalist who may be engaged in the study of the particular group of organisms of which it treats; though the several groups with which it deals have now received a much fuller exposition at the hands of modern investigators (Allman, Hincks, Gosse, &c.). Besides the work just mentioned, Dr Johnston published in 1842, 'A History of British Sponges and Lithophytes,' in which he not only dealt with the sponges properly so called, but also with a number of marine organisms which are now known to be of a vegetable nature. Owing to the exceptional difficulties which attend the study of the sponges, and the comparatively very limited knowledge possessed by the naturalists of fifty years ago as to the structure and nature of the sponges in general, this work is not of nearly so much value as the one on the British Zoophytes.

The other work alluded to above is the 'Rare and Remarkable Animals of Scotland,' by Sir John Graham Dalyell. This handsome work, published in 1847 by Van Voorst, in two quarto volumes, deals with the Cœlenterate animals or Zoophytes of the Scottish seas, and is illustrated by beautiful coloured plates. To Sir John Dalyell is due the credit of having independently worked out the extraordinary phenomena attending the production of the great swimming jellyfishes from the little fixed Trumpet-polype or Hydra-tuba—one of the most wonderful chapters in zoological history. It is true that this subject had been previously investigated successfully by the celebrated Norwegian naturalist, Sars (1829-40); but the observations of the Scottish zoologist would seem to have been made quite independently.

EDWARD FORBES.

EDWARD FORBES deserves special mention as an admirable representative of the old and honourable race of general naturalists. He was a *naturalist* in the old sense of this term, rather than a *zoologist;* and he belonged, therefore, to a genus which is in the present epoch much less largely represented than it used to be. As a matter of course, he was essentially and principally a zoologist, or an investigator of animals. He was even a *specialist* in zoology, and his name will long be remembered in connection with the British Mollusca and the British Echinoderms. But he was much more than a mere zoologist; he was an accomplished botanist, and a very able geologist.

Rarely, indeed, do we now find any one man uniting in himself high excellence in these three departments. Nor can such be well expected, in view of the enormous development that these three sciences have, one and all, undergone since the middle of this century. At the same time, there is cause for regret that specialisation should now so completely rule in all departments of natural history. Less than fifty years ago, any teacher of zoology considered a knowledge of geology and palæontology—the

latter being only a department of zoology—as an absolutely indispensable part of his equipment. At the present day, it is no very unusual thing for even a distinguished zoologist to be largely or wholly ignorant of these subjects; and in another fifty years it is more than probable that the increase in our store of knowledge will be so great, that only the exceptionally gifted will be able to master thoroughly more than a single branch of natural history. That something may be thereby gained in *depth* is probable enough; but there will be unquestionably a corresponding loss in width.

Edward Forbes was born at Douglas, in the Isle of Man, on the 12th February 1815. He was a delicate child, and received, therefore, no systematic education up to his twelfth year. He early showed a strong taste for natural history; and was one of those boys who make friends with all sorts of animals, whose pockets are always full of all sorts of beasts, birds, and minerals, who are never so happy as when in the open air, and who, naturally enough, cause as much anxiety to their teachers as does the supposititious duckling to the hen which brings it up. The few years of school-life which he ultimately had, left him with a limited amount of classical knowledge, a still smaller amount of any mathematical learning, no knowledge of any physical or natural science, and a total want of even a rudimentary acquaintance with any modern language. The one accomplishment that he had acquired was that of drawing, though it does not appear that he learned even this at school. 'Educated,' therefore, he certainly was not, when at sixteen years of age he left his Manx school for good.

It was moreover unfortunate, as Dr George Wilson, his biographer, has remarked, that 'his home-circle included no intelligent senior of his own sex, who could have wisely trained him to habits of systematic study, and taught him by precept and example the importance of rule and method in intellectual as well as physical work. For want of such training, much of his energy was unwisely directed, and he left behind him at his death a far less compact and conspicuous monument to his genius than his enormous diligence would have produced, had his intellect revolved in an orbit of smaller area, and been less liable to deflection towards new centres of attraction in every portion of its path.' At the same time, Forbes's mental idiosyncrasy was peculiar, and it is doubtful if it could have been fundamentally altered by any educational process. He had a naturally unmethodical, discursive mind. As the wise writer above quoted further remarks, 'the minds of some men are like diving-bells, with walls of opaque iron, and one small window at the top. Little light enters them, and that always in one direction. The minds of an exactly opposite class are crystal palaces, the walls all glass, and light entering in every direction. The choicest minds are intermediate in structure. They have windows to each point of the compass, besides a goodly skylight, but shaded corners abound under all degrees of illumination short of exposure to the direct glare, and there are shutters to close each window when that is desirable, and prevent the confusion of conflicting cross-lights. Edward Forbes's intellect was of the second class, and open at every moment to all the skyey influences. It would have been better in some respects if he had been persuaded in early life to make it less than all window, by a shutter here and there; but he loved the full light, and all that he could be induced to do was to temper the brightness by a veil, originally but one degree less transparent than the glass, and even when thickest, more translucent than opaque.'

In 1831, Edward Forbes, having completed his brief school career, visited London, where he stayed between three and four months. He had the intention of entering upon the study of art, in order to become a professional artist; but he had mistaken his vocation. He was refused admission to the Royal Academy as an art-student, and a well-known artist under whose tuition he had placed himself held out to him no encouragement to follow his proposed career any further. Under these circumstances Forbes very sensibly abandoned all idea of art as a profession, and made up his mind to study medicine. He therefore proceeded to Edinburgh, where he entered the university as a medical student in November 1831.

Forbes's career as a medical student cannot be touched upon here. He studied botany under Professor Graham, and natural history under Professor Jameson, both well-known men in their day. It must be remembered, however, that scientific teaching at that time was a very different thing from what it is now. It was, as Dr Wilson says, a matter of dispute among the Edinburgh students whether Professor Graham had altogether '*six* or *seven* diagrams to illustrate the structure of plants. A microscope was never seen in the class-room, and the majority of students could not have told with confidence which end of the tube should be put to the eye. No instruction was given in dissecting or examining plants, further than by pulling them to pieces with the fingers, and examining them with a pocket-lens.' In the department of natural history, things were very much the same. Jameson, the then professor of natural history, able man as he was, nevertheless was more of a mineralogist and geologist than a zoologist. The anatomical side of natural history was almost wholly neglected by him; and the university museum was almost entirely without morphological specimens of any kind, such as skeletons, dissections of animals, or even models of minute structures. No Invertebrates were to be seen in the museum, with the exception of examples of such groups as the insects, shellfish, or corals. It would appear also that Jameson was essentially a student himself, rather than a teacher.

Forbes began his systematic studies in zoology, however, at a specially favourable time. The microscope had been up till that time 'an instrument understood and handled by few, and by such was regarded with much the same feelings as an enthusiastic musician regards his Cremona violin.' Now, however, great improvements had been effected in its mechanical construction; and not only had the instrument thus gained greatly in efficiency and simplicity, but it could be produced at a price so much reduced as to render it possible for any one to purchase one. The microscope became, therefore, at this time the inseparable companion of all naturalists, the weapon of precision for all new assaults on biological problems. 'Histology,' or the science which deals with minute structures and tissues, sprung suddenly into full existence. The smaller forms of animal and vegetable life, which till now had been imperfectly understood, or had wholly defied investigators, began now to be slowly elucidated, and arranged in their proper places in the system of nature. Endless problems in physiology commenced to receive their final solution. In short, the whole face of the biological sciences underwent a rapid and fundamental alteration. .

Into all this Forbes threw himself with the utmost ardour; but his favourite studies, as was to be expected, were natural history, botany, and geology. Not only did he acquire a wide general knowledge of these subjects, so far as these were known at that time; but above all, he learned to observe and investigate for himself. Even in his second year of university study, we find him giving the preference to the more scientific branches of the medical curriculum over the more strictly technical and professional subjects. Year after year, this process of clinging more and more closely to the natural sciences went on; till in 1836 he finally and formally abandoned the study of medicine.

During his vacations Forbes had wandered over many parts of our country, and had occupied himself in all sorts of scientific, principally zoological observations, and in collecting specimens illustrative of the natural history, botany, and geology of the districts which he visited. On one occasion he extended his tour to Norway, in company with a fellow-student, and at another time he travelled through parts of France, Switzerland, and Germany. At an early period also he began dredging in the British seas, a practice which he afterwards prosecuted with such success and such brilliant scientific results. Forbes may be said, in fact, to have been one of the first British naturalists to recognise the enormous value of the dredge as an instrument of zoological research; and, from this time on, we find him engaged in dredging whenever he got an opportunity. The results of his dredging expeditions round our coasts were given to the world in various memoirs, the first of which was published in the 'Annals and Magazine of Natural History' as early as 1835, when he was still an Edinburgh student.

In 1836, Forbes, having finally renounced medicine, proceeded to Paris, where he stayed till the following year, studying natural history under Geoffroy St Hilaire and De Blainville, and working in the great museums of the French capital. At the close of the Paris session he paid a visit to the south of France, and from there he made his way to Algeria, where he collected a number of Molluscs, which he subsequently described in the 'Annals of Natural History.' The winter of 1837-38 Forbes again spent in Edinburgh, nominally as a literary student, but in reality he worked at nothing but science. It was at this time that he published his first book, a little treatise entitled 'Malacologia Monensis,' dealing with the Mollusca of the Isle of Man. The summer of 1838 was spent once more on the Continent, and the winter of the same year found him back again in Edinburgh—always hard at work writing papers and scientific memoirs, giving lectures on natural history, collecting, and the like. The summer of 1839 was spent mostly in dredging round the coasts of Scotland, and in collecting materials for a report on the air-breathing

Mollusca of Britain for the British Association. This report Forbes laid, together with some other zoological papers, before the meeting of the Association in the same autumn; and at the same meeting he founded the celebrated dinner of the 'Red Lions.'* The whole of 1840 was spent in scientific work, a great part of which was directed to the preparation and publication of his well-known treatise on the sea-urchins, starfishes, and other Echinoderms of Britain.

In the meanwhile Forbes, though steadily gaining reputation, had failed to obtain any fixed employment in science. Lecturing had proved pecuniarily a failure, and his scientific works did not bring him in any money. In the early part of 1841, however, he was offered the post of naturalist to the surveying-ship *Beacon*, which was about to start on an expedition to the Levant. This offer he accepted, and after a fruitless attempt to obtain the natural history chair in Aberdeen—to which Macgillivray was appointed—he started for the East on board the *Beacon*, in company with his friend and fellow-naturalist, William Thompson. It is not necessary to enter into any details as to the incidents of the expedition, which occupied two years. Not only was Forbes enabled to make a number of interesting and important observations as to the distribution of marine animals in the Mediterranean, but he was also able to spend some three months in a tour in Asia

* The dinner of the 'Red Lion Club' was founded by Forbes in 1839, at the meeting of the British Association at Birmingham. The name was derived from the tavern at which the meeting of the club took place; and the dinner became an annual feature, which is still kept up at every British Association meeting. At the dinner of the 'Red Lions,' the learned guests are supposed to drop all their science, and to give themselves up wholly to fun and merriment, approbation of the songs or speeches being expressed by roars and growls.

Minor. A narrative of this, written in conjunction with his friend and companion, Captain (then Lieutenant) Spratt, was subsequently published under the title of 'Travels in Lycia' (1847). As regards the scientific results of this expedition, by far the greatest interest attached to the researches which Forbes at this time carried out as to the distribution of the Shellfish and Radiate animals at different depths in the sea. This had long been a favourite subject with him, and he had previously begun to divide the British seas into 'zones' of different depths, characterised by particular assemblages of animals. It will, however, be best to defer consideration of this subject till the completion of this brief sketch of Forbes's life.

At the close of the year 1842, Forbes returned to England, when he found that he had been in his absence appointed Professor of Botany in King's College, London. This appointment he gladly accepted, as his father had met with pecuniary losses, and was no longer in a position to help him. Two or three months after his return, he was also appointed Curator of the Museum of the Geological Society of London. He was thus plunged into a constant whirl of work, principally of the thankless official kind, and he found comparatively little time for original research. Not only were his duties numerous and trying; but the emoluments of his combined offices of professor and curator did not bring him in much more than about £200 per annum, and he was thus forced to do literary work of the 'pot-boiling' kind.

In 1844, however, his position was somewhat ameliorated by his being appointed to the newly created post of Palæontologist to the Geological Survey, whereby he was enabled to resign the curatorship of the Geological Society. He found himself now in a much more congenial sphere. His connection with the Geological Society had strengthened his early fondness for geology. He now had the opportunity—indeed it was now his duty—to enter fully into the study of palæontology, one of the great charms of which is that, though essentially a part of zoology, it can hardly be successfully approached save through the avenue of geology, while, in one of its subordinate aspects, it really forms a department of geological science. Besides, Forbes had always had a special interest in all questions affecting the 'distribution' of animals, and there is no department of natural history more fruitful in problems of this kind than palæontology.

The remainder of Forbes's too short life may be told in a very few words. The next few years were spent in constant work of all kinds—dredging, geologising, palæontologising (to coin a much-needed word), lecturing, and above all, writing incessantly. His personal relations with his colleagues on the Geological Survey were of the happiest kind. His reputation in scientific circles was of the highest. His work was, much of it, thoroughly congenial. He had little to complain of beyond the fact that, in accordance with the traditional treatment of science and of higher learning generally in Britain, he was greatly overdriven, and was so ill paid that he was compelled to do hack-work of various kinds in order to exist. The primary result of this short-sighted policy was that one of the finest and most original minds Britain has produced in the last half-century was forced to expend a large part of its energy in mere drudgery that could have been as well, or perhaps better, done by one of meaner capacity. A secondary result, it need not be doubted, was the impairment of his health and the shortening of his term of life.

In 1848 Forbes married, his wife being the daughter of General Sir C. Ashworth. In 1851, the Royal School of Mines, in connection with the Geological Survey, was founded, and Forbes was appointed to the professorship of natural history in the new institution. During his tenure of office the School of Mines never became prosperous or popular, and its comparative failure was a source of great disappointment to him. In 1853 Professor Jameson at last resigned the chair of natural history in the university of Edinburgh, towards which Forbes had been looking for many years. To this chair, after some delays, Forbes was ultimately appointed, thus realising what had been the ambition of his life. The relief from the harassing overwork of years had at last come to him; but it had unfortunately come, as it proved, too late. In the summer of 1854, Forbes gave his first course of lectures in his new chair, to a very large class, and with brilliant success. In the early part of this year he had been elected to the highly honourable position of President of the Geological Society of London; and in the autumn of this year he filled the presidential chair in the geological section of the British Association. On the 1st of November he delivered the introductory lecture to the class of natural history, but after a few days of lecturing he was attacked by a severe illness, which from the first assumed a very serious aspect, and to which he succumbed on the 23d of the month. Thus died one of the most accomplished and original naturalists that this country has yet produced, in the prime of life, for he was not forty years old, in the zenith of his fame, and at the moment when he had just commenced the happiest and most hopeful period of his laborious career.

Of Forbes's multifarious scientific publications but few can be noticed here. His special subjects in natural history proper had always been the Zoophytes, the Echinoderms, and the Mollusca, together with all problems relating to the geographical distribution of animals. As regards the first of these subjects, his chief contributions are the following:

(1) 'On the Morphology of the reproductive system of the Sertularian Zoophyte.' This memoir was published in the 'Annals and Magazine of Natural History' in 1844, and in it he demonstrated that the singular urn-shaped capsules which are seasonally produced by the common Sertularian Zoophytes of our seas, and which have the function of reproducing the species, are really a modified condition of the ordinary buds of the colony, the function of which is nutritive. In other words, he showed that just as the flowers of a plant are only specially modified buds, and therefore composed of altered leaves; so the reproductive buds of the sea-firs and their allies are only modifications of the ordinary nutritive 'polypites.'

(2) 'On the Pulmograde Medusæ of the British Seas,' published in the 'Annals and Magazine of Natural History' in 1846.

(3) 'A Monograph of the Naked-eye Medusæ,' published by the Ray Society in 1848. In this last well-known

work Forbes described and figured all the recognised species of the smaller British jelly-fishes, which are called 'naked-eyed,' because the little coloured eye-spots are placed conspicuously round the margin of the swimming-bell.

With regard to the star-fishes, sea-urchins, and other Echinoderms, Forbes wrote a number of valuable works, to say nothing of detached memoirs in different scientific periodicals. The chief of these are:

(1) 'A History of British Star-fishes and other Animals of the class Echinodermata.' This familiar work was published by Van Voorst in 1841 as one of his series of treatises on British zoology. It not only contains illustrations of all the species described, but is embellished with the picturesque or fanciful tailpieces which Forbes loved to design.

(2) A 'Monograph of the British Fossil Asteriadæ,' published in the memoirs of the Geological Survey in 1848.

(3) A 'Monograph of the Silurian Cystideæ of Britain,' also published in the memoirs of the Geological Survey, in the same volume as the preceding.

(4) 'Figures and Descriptions of British Organic Remains: Echinoderms.' This constituted the third 'decade' of a series of palæontological publications issued by the Geological Survey; the title of 'decades' being given to them because each number was supposed to contain ten plates.

(5) 'A Monograph of the British Tertiary Echinoderms,' published by the Palæontographical Society in 1852.

Thirdly, in the department of the Shellfish (*Mollusca*),

Forbes was the author of a number of very valuable memoirs, descriptive of species, or dealing with the distribution of these animals, and treating not only of the marine types, but also of the land-shells and the freshwater forms. In this branch, however, his great work was his 'History of British Mollusca,' written conjointly with Mr Hanley. This standard work was in four large volumes, the first of which appeared in 1848.

Much of Forbes's work as regards the Echinoderms and the Molluscs was of a palæontological character, and dealt with fossil species. Apart from this, he did some excellent work in what is called 'stratigraphical' palæontology, or in other words in palæontology as *applied* to geology. His two best-known memoirs in this connection are one 'On the Succession of Life in the Dorsetshire Purbecks,' and one 'On the Fluvio-marine Tertiaries of the Isle of Wight.' The first of these was published as one of the 'Reports' of the British Association in 1850; and the second appeared in the Quarterly Journal of the Geological Society in 1853.

Admirable as was Forbes's work in the various branches of natural history above enumerated, he is perhaps best known by his researches into the complicated problems connected with the 'distribution' of animals, both in the sea and upon the land. Before dealing briefly with these researches, it may, however, be well to glance for a moment at the views which he held with regard to the nature of 'species.' Now, Forbes, like almost all naturalists at that time, was a firm believer in the fixity of species. Lamarck's views as to the mutability of animal species had at this period obtained no acceptance in the scientific world. Forbes therefore, like Cuvier before him, believed that each 'species' of animals and plants was so far permanent, that though it might be exterminated altogether, and thus become 'extinct' it could not become changed into a new species. This implies the further belief that 'variation' is strictly limited and definite in amount, and that 'varieties' of animals are mere temporary modifications instead of being 'incipient species.' To use Forbes's own words, 'every true species presents in its individuals, certain features, *specific characters*, which distinguish it from every other species; as if the Creator had set an exclusive mark or seal on each type.' He also believed that what we call a 'genus,' that is to say, a group of allied species, is similarly permanent. He believed— to use his own expression—that 'a genus is an abstraction, a divine idea .... a true genus is natural, and, as such, is not dependent on man's will.'

This belief in the fixity of species carries with it, as an almost necessary corollary, a belief in what are known as 'specific centres.' On the view of the permanence of species, each particular species must have come into existence at a particular moment of time and at a particular place in space. That place must have been the point where the first progenitor, or pair of progenitors, of the species was created. This place must be supposed to be one specially adapted for the life of the species, and it constitutes the 'specific centre' for the species. From this centre the species would gradually diffuse itself by migration, over a more or less extensive area, till its further progress would be stopped by meeting conditions unsuitable for its existence. Hence, each species at the present day is found inhabiting a larger or smaller 'specific area;' and there is found in that area one point—the 'metropolis' of the species—where the individuals are more abundant than elsewhere, and which may therefore be taken as the point where the species was originally created. As a matter of course, no species could have more than one 'specific centre.' If, therefore, a species should be met with in two quite detached areas—as sometimes happens—this must be explained on the supposition that the original area of the species had become divided into two in consequence of changes in the physical geography of the area. Or, it might be supposed that some individual of the species had been accidentally transported from its original area to some new place, where the conditions happened to be suitable for its existence and propagation.

Forbes further believed that when a species had once become extinct, it was never re-created. We sometimes find, however, that a given species, after living a long time in some particular region, disappears altogether from that area, and that, after a longer or shorter period, it reappears again in the same place. This phenomenon was explained by Forbes on the supposition that the species had been forced to abandon its original area, in consequence of some change of conditions which rendered its further existence there impossible, and that it had therefore migrated to some adjoining area where it met with suitable surroundings. At a later period, however, the conditions of the original area might again become favourable to the species, and then it would migrate back again. In this case, therefore, there is no real extinction of the species, but only its disappearance temporarily from its original area.

Lastly, Forbes considered that 'genera,' or natural groups of species, are distributed very much as species are, each genus having its 'generic area.' When a genus includes a large number of species, there may be found within the generic area 'a point of maximum (*metropolis*) around which the number of species becomes less and less. A genus may have more centres than one. It may have had unbroken extension at one time, and yet, in the course of time and change, may have its centre so broken up that there shall appear to be outlying points. When, however, the history of a natural genus shall have been traced equally through its extension in *time* and *space*, it is not impossible that the area, considered in the abstract, will be found to be necessarily unique.'

Forbes, in addition, clearly formulated what he termed the 'law of representation' among species. He showed, namely, that in all regions, however widely removed from one another, species or groups of species of animals are found which are very like each other, provided only that the conditions of these regions are similar as regards climate. In other words, wherever similar life-conditions prevail, similar species of animals will be found. In all such cases, however, though the species of such regions are *similar*, they are not *identical*. Such species he termed 'representative species.'

As, in Forbes's view, species are permanent and immutable, the only explanation which he could give of the existence of these representative forms was that similar but specifically distinct types of animals and plants had been created in all regions in which the conditions of life were the same. Of the *fact* of the existence of these 'representative' groups of species, or of representative species, no doubt can be entertained; but modern naturalists would explain their origin otherwise. At the present day it would not be admitted that two representative species had been specially created in two areas where similar conditions prevailed. On the contrary, the modern view would be, that two representative species owe their *likeness* to the fact that they are the descendants of a common ancestor; and that their *unlikeness* is due to the fact that, having become widely separated by migration, and kept separate by the formation of some natural barrier, they have gradually become modified by variation, till we now speak of them as distinct species.

Some of the most interesting points connected with Forbes's researches into the distribution of animal life relate to the distribution of animals in the sea. Many of these researches were carried on in his dredging excursions round the British coasts; but his visit to the Mediterranean enabled him to carry out a series of elaborate investigations on a more extended scale, and at greater depths than he had previously explored. These investigations were published by Forbes in his well-known 'Report on the Mollusca and Radiata of the Ægean Sea,' which he laid before the meeting of the British Association at Cork, in 1843. As regards British seas, and more or less markedly round all coast-lines in all regions, Forbes recognised four very well defined 'zones of depth,' each characterised by particular types of animals.

The first of these zones is the so-called 'littoral zone,' embracing the tract between tidemarks, and characterised, not only by its abundant development of different kinds of seaweeds, but also by a number of peculiar animals. The littoral animals must be capable of being left uncovered twice a day by the receding tide, and they must also be able to withstand exposure to the direct rays of the sun. Hence, the animals of the littoral zone are mostly referable to peculiar types; and the same types, or 'representative' types, are found between tide-marks in almost all parts of the world.

Below low-water mark, and extending to the depth of about fifteen fathoms, is what is known as the 'laminarian zone,' so called from the abundance in it of the great strap-like 'tangle' (*Laminaria*). The fronds of this seaweed form a sort of submarine forest in the shallow water all round our coasts, the upper edge of which is just visible at the lowest ebb of the spring-tides; and it affords shelter and food to a vast abundance of marine animals. Many of the species of the laminarian zone are peculiar, and they are often remarkable for the brilliancy of their coloration.

A third zone was termed by Forbes the 'coralline zone,' and it extends from about fifteen fathoms to fifty fathoms in depth. In this zone, plants are chiefly represented by the peculiar coral-like calcareous *Algæ* known as the 'Corallines' and 'Nullipores.' Animal life is extremely abundant, and many of the species are peculiar.

Lastly, Forbes recognised a fourth or 'deep-sea coral

zone,' extending from fifty fathoms to about one hundred fathoms or more in depth. In this zone animal life is much less abundant than in the preceding, and among the more characteristic types—in the northern seas, at any rate—are certain kinds of corals, which only live and flourish in deep water.

In his researches in the Ægean Sea, Forbes recognised the occurrence of similar zones, but he thought they were capable of further subdivision. He recognised in the Eastern Mediterranean, in all, eight well-marked 'regions of depth,' each of which is 'characterised by its peculiar fauna, and when there are plants, by its flora.' The lowest of these zones was beyond the limit of what he had called the 'deep-sea coral zone,' and extended from a hundred and five fathoms in depth to two hundred and thirty fathoms, beyond which depth his explorations were not carried. In this zone, animal life was extremely sparse; the species were mostly small; and the shells were mostly pale-coloured or white, in part, apparently, owing to a deficient supply of light. As just remarked, Forbes did not dredge at a depth of more than two hundred and thirty fathoms, which at that time was considered an extraordinary depth for the carrying out of dredging operations. Owing to the great reduction in the number of species of animals which he found at this depth, he came to the conclusion that at depths greater than this the animals would become fewer and fewer, and that they would ultimately be found to disappear altogether. In accordance with these observations, he placed the 'zero of animal life' at about three hundred fathoms, and he concluded that at all depths greater than this animal life would be found to be altogether wanting.

Forbes's conclusions on this point met with the general assent of naturalists, and it became an accepted doctrine in zoology that animal life was wanting in the deeper portions of the ocean. It would be out of place here to detail the various steps which have led to an entire reversal of Forbes's dictum on this point. Isolated observations, both before and after Forbes's time, had been recorded, which indicated the existence of animals at great depths in the sea; but these had been neglected by naturalists, or had been looked upon with suspicion. At present, however, the interest in these detached and solitary observations is chiefly of an historical kind. Of late years various nations, and notably our own, have caused extensive and systematic explorations to be made as to the physical and biological conditions prevailing in the larger oceans. By various specially-equipped expeditions—the most important of which was that of the *Challenger*—the bottom of the deep sea has been systematically explored by the dredge and trawl up to depths of between three and four thousand fathoms, upon a scale and with a completeness entirely beyond the reach of any private individual. We now know that there is no 'zero of animal life' in the sea. So far from animals ceasing to exist at depths greater than three hundred fathoms, they are found at all depths in the sea. Nor are the animals of the 'abyssal zone' less numerous than those of shallower waters. On the contrary, animal life exists in abundance even at the greatest depths, where the temperature is suitable; and it is the *temperature* of the water at the bottom, rather than the *depth* of the water, that is the predominant factor in determining the nature and the plentifulness of the forms of animals with which the floor of 'the deep sea' is peopled.

In addition to the investigations which he carried out into the present distribution of animals and plants, and particularly the laws which govern the occurrence of animals at different depths in the sea, Forbes interested himself greatly as to the *causes* of the presence of particular types of animal or vegetable life in special regions. As regards this subject, he was greatly assisted by his wide knowledge both of the palæontological department of natural history and also of geology proper. Various of his published writings deal more or less extensively with this problem; but the one most generally known is his famous memoir 'On the Connection between the Distribution of the existing Fauna and Flora of the British Isles, and the geological changes which have affected their area, especially during the epoch of the Northern Drift' ('Memoirs of the Geological Survey of Great Britain,' vol. i., 1846). This memoir affords such an admirable example of the way in which Forbes combined and brought into a focus his varied knowledge of zoology, palæontology, and geology, that it may be well to glance for a moment at some of the results at which he arrived. In so doing, however, it will be best to leave the animals altogether out of sight, and to restrict our attention entirely to the conclusions which Forbes reached as to the origin of the existing British plants, the problem as regards these being of a less complex character than it is in the case of the former.

It is to be remembered, to begin with, that in dealing with this problem, Forbes started with a firm belief in the doctrine of 'specific centres,' or geographical points from which the individuals of each species have slowly diffused themselves. The problem before Forbes, therefore, was how to account, on the basis of 'specific centres,' for the present distribution of animals and plants in Britain. As regards the plants, which alone we shall consider here, Forbes showed that there exist in Britain certain well-marked areas or regions, characterised by peculiar types which do not occur elsewhere. Of these botanical provinces he distinguished five, as follows :

(I.) An area which may be spoken of as the 'Asturian area,' comprising the mountainous districts of the west and south-west of Ireland. Here we meet with a number of peculiar plants, comprising the Arbutus, the Mediterranean Heath (*Erica mediterranea*), and several peculiar species of saxifrages. The nearest point in Europe where the same plants are found growing as natives is in the Asturias, in the north of Spain.

(II.) A second area comprises the south-west of England (Devonshire and Cornwall principally), together with the south-east of Ireland. In this area we find a number of plants of what may be called the 'Armorican' type, unlike those found in Britain generally, but intimately related to the plants of the Channel Islands and of Brittany and Normandy. Amongst these are the Cornish Heath (*Erica vagans*), the Wild Madder (*Rubia peregrina*), the French Tamarisk (*Tamarix gallica*), and various other peculiar forms.

(III.) A third area comprises the south-east of England,

where the plants correspond in many respects to those of the opposite coast of France, characteristic forms being the Sainfoin (*Onobrychis sativa*), the Bryony (*Bryonia dioica*), several species of Mullein (*Verbascum*), and the Box (*Buxus sempervirens*). This is the least well defined of the areas which Forbes distinguishes, though many of the land-snails which it possesses are peculiar. Both as regards the plants and the animals, however, the species are types which are especially fond of districts where the white chalk is the underlying formation, or where at any rate the soil is calcareous. The peculiarities of their distribution, therefore, depend essentially upon the distribution of a soil suitable for them.

(IV.) A fourth area may be distinguished by the name of the Scandinavian or Arctic area, and comprises the Highlands of Scotland, the higher parts of the Lake District of Cumberland and Westmorland, and the more elevated parts of Wales. Here we meet with numerous species of plants identical with, or closely allied to, forms which are characteristic of Scandinavia or of the Arctic regions. Thus, we find the Blue Gentian (*Gentiana nivalis*), the Dwarf Birch (*Betula nana*), the Scotch Primrose (*Primula scotica*), the Alpine Veronica (*Veronica alpina*), two Dwarf Willows (*Salix herbacea* and *S. reticulata*), and other familiar northern types. There is, however, a progressive diminution of these alpine forms of plants as we proceed southwards; the largest number of them being found in the Highlands, a smaller number in the Lake District, and a still smaller number in Wales.

(V.) The last area comprises all parts of the British Isles not enumerated in the four regions previously

216   NATURAL HISTORY.

mentioned, and it is peopled by the general flora of our country, which is everywhere present, alone or in company with the others. The plants of this area are

Map showing the Distribution of Plants in Britain. (After Forbes.)

identical with those of Central and Western Europe, and may therefore be spoken of as forming the 'Germanic type' of vegetation. This *general* flora of Britain comprises such universally distributed types as the daisy, the primrose, the buttercup, the lesser celandine, and, in fact, our ordinary flowering-plants, together with our common shrubs and trees. Some of the less abundant species of this general flora are confined to the eastern counties of England, and a considerable number of common English types are not found to occur in Ireland.

The question next arises—how can the above remarkable facts as to the distribution of plants in Britain be accounted for? In attempting the solution of this problem, Forbes points out three modes in which an isolated area (such as the British Islands) might become in the first place peopled by plants or animals:

(1) By special creation within the area.
(2) By transport to it.
(3) By migration before isolation.

The first of these modes needs no consideration, as the British animals and plants, taken as a whole, are identical with those of the continent of Europe, and there is therefore no necessity, and indeed no room, for the supposition that they were specially created for our area.

The second mode, namely introduction to the region by transport, is insufficient. Supposing Britain to be, as it now is, separated from the Continent, artificial or natural means of transport might doubtless serve to explain the existence with us of certain animals and plants; but it assuredly would not adequately account for the distribution of the general flora and fauna of our country.

The third hypothesis, therefore, alone remains—namely, that our animals and plants have been introduced *by migration before isolation.* In other words, our present animals and plants have, in the main, simply reached us by migrating hither at a time when Britain was not an island, but was directly connected with other regions by land. As regards the general or 'Germanic' flora, there is no difficulty whatever in accepting this theory. We have the most abundant evidence that at a geologically very recent period (subsequent to the coming into existence of our ordinary animals and plants), Britain was directly connected with the continent of Europe, the English Channel and the German Ocean being in part or wholly converted into dry land. During the same period of elevation, Ireland was united with England, by the obliteration of the intervening sea. It was, then, during the continuance of this land-connection, that all our generally distributed plants and animals migrated to our area from the Germanic regions of the Continent. As this migration took place from the east, the Germanic animals and plants necessarily reached England at an earlier time than they found their way to Ireland. We may further explain the absence of certain common English plants and animals (as, for example, snakes) from Ireland upon the supposition that the ultimate separation of Ireland from England took place prior to the severance of the latter country from the Continent.

Modern geological researches also enable us to explain, without any difficulty, the existence in the Highlands of Scotland and the mountainous districts of the north of England and of Wales, of plants peculiar to Scandinavia or to the Arctic regions. We know, namely, that during the glacial period, the greater part of our islands participated in the frigid and Arctic conditions which prevailed in Northern Europe, and in the northern portions of North America. At this time, our mountains were covered with ice and snow, and the general state of things must have been very much what we now see in the Arctic regions. During this period of glacial cold, a great migration southwards of Arctic animals and plants took place, and these, finding suitable life-conditions, established themselves far south of their former limits. Thus, during this period the Icelandic Scallop (*Pecten islandicus*), and various other shellfish which are now found living in Arctic seas, migrated as far as the coasts of Britain, and we find their remains in the glacial clays of the estuary of the Clyde, and elsewhere round our coasts. Similarly various alpine and Arctic plants invaded our area, and took advantage of the cold to establish themselves on the low grounds and hills. Contemporaneously with this southward migration of northern plants and animals, we must suppose a like movement to have taken place on the part of the animals and plants which had previously inhabited our area. These latter would be driven gradually farther and farther south by the increasing cold, and a clear field would in this way be left for the Arctic invaders.

When, however, the glacial period ended, and more temperate conditions were gradually re-established, a reverse movement would be set on foot. The southern forms of life would again move northwards, and piece by piece reconquer the territory from which they had been dispossessed; while the northern immigrants would be driven, step by step, backwards towards the pole. But as the climatic conditions became gradually less severe, many of the Arctic animals and plants (the latter more especially) would not retire northwards, but would be driven from the low grounds to the more mountainous parts of the country, when the temperature would still be cold enough to suit them. Some of these, moreover, would in this way succeed in maintaining a permanent foothold in the country, since the elevation above the sea-level to which they had retired, would secure them a sufficiently low temperature for their existence. The above processes must, of course, have been very slowly effected; but we need not doubt that Forbes's views on this point were correct, and that the Scandinavian and Arctic plants now living in the Scottish Highlands, in the Lake District, and in Wales, are only the survivors of a much greater number of northern types of life which invaded us during the cold of the glacial period. We can also easily understand, on this view, how it should be that the Highlands, lying as they do nearer to the original home of the northern invaders than either Cumberland or Wales, should now possess a greater number of these Arctic species than do the two districts last named.

There remain for consideration, the three smaller floras which Forbes distinguished as occurring in the British area. The plants characteristic of these three floras are, according to Forbes, 'derived assemblages of plants south of the great Germanic group. As the south of England and of Ireland were in all probability unsubmerged during the glacial epoch, they may have come over either before, or during, or after that epoch. There are strong reasons for believing they migrated before.'

The plants of the south-east of England, constituting

what Forbes called the 'Kentish Flora,' must have been derived from the north-western provinces of France, and must have migrated into our area at a time when the Strait of Dover had no existence. We have no evidence as to the precise time when the Strait of Dover was formed; but it is not improbable that its formation, as believed by Forbes, was anterior to the severance of the general land-connection between the eastern counties of England and the opposite shores of Belgium and Holland. In this case the Kentish flora would be considerably older than the general 'Germanic' flora of our country.

The plants of the second or 'Armorican' flora must have migrated into Devonshire, Cornwall, and South-eastern Ireland, at a time when all these regions were connected with one another and also with Brittany and Normandy by land. According to Forbes's view, this must have taken place at a time anterior to the great glacial submergence of Britain; so that the Armorican flora is also more ancient than either the Scandinavian or the Germanic floras.

With regard to the Asturian flora of the south-west of Ireland there are greater difficulties. The plants of this flora are species which 'at present are forms peculiar to or abundant in the great peninsula of Spain and Portugal, and especially in Asturias.' There are, however, many grounds for believing that these plants migrated from Spain into Ireland at a period when there existed a direct land-connection between these two regions, now separated by such a wide stretch of sea. Difficult as it may appear to establish any reasonable probability of there having existed (since the differentiation of our existing species of plants took place) a connection by continuous land between Spain and Ireland, Forbes boldly faced the problem. He brought forward geological and palæontological evidence in support of the daring hypothesis that towards the close of the Miocene period a great tract of land, 'bearing the peculiar flora and fauna of the type now known as Mediterranean, extended far into the Atlantic —past the Azores—and that, in all probability, the great semicircular belt of Gulf-weed ranging between the fifteenth and forty-fifth degrees of north latitude, and constant in its place, marks the position of that ancient land, and had its parentage on its solid bounds.* Over this land that flora, of which we have now a few fragments in the west of Ireland, might with facility have migrated.' On this hypothesis, the peculiar 'Asturian' plants of Cornwall, Devonshire, and Western Ireland are the remains of the oldest flora in the British Islands, and their introduction into our area took place at the end of the Miocene period.

* The Gulf-weed (*Sargassum bacciferum*) is the seaweed which gives rise to the 'Sargasso-sea' so well known to navigators since the time of Columbus. Though not now attached, it is very closely related to species of *Sargassum* which are essentially littoral seaweeds, or live in shallow water near the shore. Eminent botanical authorities, therefore, are of opinion that the Gulf-weed was at one time a fixed seaweed, and that its present condition is an abnormal one. As the present Gulf-weed does not propagate itself by fructification, but apparently simply by breakage, this view would seem to be very probably correct. In this case there is much to be said for the hypothesis of Forbes, that the present belt of Gulf-weed in the Atlantic marks the position of an ancient coast-line, now deeply submerged.

THE DAWN OF THE EVOLUTIONARY PERIOD.

ERASMUS DARWIN.

WE enter now upon the final phase of zoological science, so far as we are here concerned—namely, the phase in which naturalists definitely accepted the principle of 'Evolution' as the key to biological problems of all sorts, and more especially as explaining the much-vexed question of the origin of 'species.' The two names which are most intimately associated with the modern theory of the origin of 'species' by 'Descent with Modification'— the 'Descendenz-theorie' of the Germans—are those of Lamarck and of Charles Darwin. The former of these wrote his 'Philosophie Zoologique' in the beginning of the present century; and the latter gave to the world his epoch-making work on 'The Origin of Species by means of Natural Selection' in the year 1859. The views put forward in these two celebrated works will be shortly sketched hereafter. In order, however, to trace with any pretence to completeness the beginnings of the modern theories as to the evolution of living beings, it is necessary to return to the pre-Cuvierian period, and to glance at the life and scientific opinions of Erasmus Darwin, the grandfather of Charles Darwin, who may be regarded as the first expositor, in any systematic form, of the doctrine of the evolution of plants and animals from pre-existent species.

Erasmus Darwin was born on the 12th of December 1731, at Elton Hall, near Newark, Nottinghamshire. He was educated as a medical man, and had a most successful professional career, first in Nottingham, then in Lichfield, and finally in Derby. He married twice; and his son, Robert Waring Darwin, the father of Charles Darwin, was the offspring of his first marriage. He died in 1802.* To quote Mr Grant Allen : † 'A powerful, robust, athletic man, in florid health and of temperate habits, yet with the full-blooded tendency of the eighteenth century vividly displayed in his ample face and broad features, Erasmus Darwin bubbled over with irrepressible vivacity, the outward and visible sign of that overflowing energy which forms everywhere one of the most marked determining conditions of high genius. Strong in body and strong in mind, a teetotaler before teetotalism, an abolitionist before the anti-slavery movement, he had a great contempt for weaknesses and prejudices of every sort, and he rose far superior to the age in which he lived in breadth of view and freedom from preconceptions.'

From very early years Erasmus Darwin had shown a

---

* An account of Erasmus Darwin is to be found in Dr Krause's 'Erasmus Darwin und seine Stellung in der Geschichte der Descendenz-theorie,' 1880. Further details are given in the same author's 'Charles Darwin und sein Verhältniss zu Deutschland,' 1885.

† Grant Allen, 'Life of Charles Darwin,' 1885.

strong leaning towards poetry, which in later life he was enabled to indulge, not altogether to the advantage of his reputation. To quote Mr Grant Allen once more, his poetry, 'though ingenious as everything else he did, had a certain false gallop of verse about it, which has doomed it to become, since Canning's parody,* a sort of warning beacon against the worst faults of the post-Augustan decadence in the ten-syllabled metre.' Erasmus Darwin's poetical works, however, though not worthy of preservation as specimens of poetical art, derive a historical interest from the scientific conceptions which they embody. The best known of them is the curious treatise entitled 'The Botanic Garden.' The second part of this singular lucubration appeared anonymously in 1788, under the name of 'The Loves of the Plants;' the first part, entitled 'The Economy of Vegetation,' not having been published till 1790. 'The Botanic Garden' dealt, in poetical fashion, with the life of plants, and it was at first received with much favour by the public, though its popularity was short-lived, and at the present day it is probably never read at all.

Erasmus Darwin's most famous and really most important work was, however, his 'Zoonomia, or the Laws of Organic Life,' the first edition of which was published in London (1794-96), in two volumes quarto. The special interest of this work lies in the fact that in it can be traced the foreshadowing of a large portion of the modern theory of the evolution of living beings. In so far, therefore, as this is the case, it may be fairly claimed

* Canning wrote a parody upon Erasmus Darwin's 'Loves of the Plants' with the title 'Loves of the Triangles.'

that to Erasmus Darwin, rather than to Lamarck, belongs the honour of having first given coherent expression to those vague ideas as to the origin of species from pre-existent species, which, floating formlessly in the minds of many of the thinkers of this period, ultimately crystallised into the modern theory of 'Descent with Modification.' It is only, however, when read in the light of our present knowledge that the real value of Erasmus Darwin's 'Zoonomia' becomes evident, and that we can recognise how greatly it was, as regards some of its leading ideas, in advance of the time at which it was written. In this respect, as remarked by Dr Krause, Erasmus Darwin suffered a fate similar to that of Goethe, 'in whose prophetic glances into the world of science the experts of the day would see nothing but the melancholy consequences of dilettanteism.'

It would be out of place here to enter into a detailed analysis of a work so complex, and in many respects so miscellaneous, as the 'Zoonomia;' but it may be of interest to indicate to what extent it contained the germs of the modern theories of evolution. It is to be remembered that Erasmus Darwin lived and wrote at a time when the great majority of naturalists believed implicitly in the doctrine of the immutability of 'species.' A few writers had ventured to suggest the possibility of the transmutation of species, but mostly in doubtful fashion, or upon purely speculative grounds. Buffon, about the middle of the eighteenth century, had clearly hinted in his 'Histoire Naturelle' at the possible, or even probable, evolution of species from pre-existing species; but he had so carefully hedged upon the point, that his opinion was deprived of almost all the weight which might otherwise have been attached to it. Goethe, likewise, arrived at the idea of the mutability of species, but he only expressed his views 'aphoristically,' and they met with no acceptance from the world at large. Erasmus Darwin, however, firmly grasped and clearly laid down many of the principles which are involved in the modern theories of the evolution of species. His views on this question are mostly contained in the section of the 'Zoonomia' which deals with the function of reproduction; but most of the ideas to which he had been led had been more or less explicitly propounded in the previously published 'Loves of the Plants.'

In the first place, he not only recognised the natural variations which present themselves in different individuals of a species, but also those which are the result of artificial or accidental cultivation. Thus, he pointed out the numerous structural peculiarities which have been induced in special breeds of such animals as horses and dogs, which have been long exercised for particular purposes; and he drew attention to the fact that some of our domestic animals had undergone changes so great that it was now no longer possible to determine with certainty from what wild species they had their origin. He also indicated that variations, sometimes of considerable importance, such as the presence of an additional digit, or the want of the tail, might occur, and might become permanent.

In the second place, he clearly recognised the principle of heredity; and he pointed out that not only might the structural peculiarities of individual animals be transmitted to their offspring, but that even some of the *habits* of the parents may be similarly handed on to the young. It is only, he remarks, from the imperfection of language that we speak of a young organism as being a *new* animal. The young animal is 'in truth a branch or elongation of the parent; since a part of the embryon-animal is, or was, a part of the parent; and therefore in strict language it cannot be said to be entirely *new* at the time of its production.'

In the third place, he divined that the community of fundamental *structure* which can be shown to underlie the differences which separate different groups of animals, affords an *à priori* presumption in favour of a community of *descent* for these groups. On this point, he remarks: 'When we revolve in our minds the great similarity of structure which obtains in all the warm-blooded animals, as well quadrupeds, birds, and amphibious animals, as in mankind; from the mouse and bat to the elephant and whale; one is led to conclude that they have been alike produced from a similar living filament.'

Up to this point, then, Erasmus Darwin had obviously grasped several of the leading principles in the modern theory of the Origin of Species by 'descent with modification.' He understood the principles of variation and inheritance, and he comprehended the importance of 'homologous' structures as proving blood-relationship. Some of his further views, however, were more akin to those afterwards put forth by Lamarck than to those which were expounded by his illustrious grandson in the 'Origin of Species by means of Natural Selection.' Thus, he seems to have thought that interbreeding between different species of animals was a probable cause of modification. Again, he appears to ascribe to the *desires* of the individual a greater power in producing modifications of its structure than modern zoologists would be disposed to allow. On this subject his views are unfortunately not as unequivocal as might be wished, since, though he uses the word 'desires,' it is not clear that he does not really mean the *needs* of the animal, in which case he would only mean what Lamarck in his writings understood by the term 'besoins.' Erasmus Darwin's views on this point are so interesting and remarkable that we may quote in full the passage in which they are set forth :

'From their first rudiment, or primordium, to the termination of their lives, all animals,' he remarks, 'undergo perpetual transformations; which are in part produced by their own exertions in consequence of their desires and aversions, of their pleasures and their pains, or of imitations, or of associations; and many of these acquired forms or propensities are transmitted to their posterity.

'As air and water are supplied to animals in sufficient profusion, the three great objects of desire which have changed the forms of many animals by their exertions to gratify them, are those of love, hunger, and security. A great want of one part of the animal world has consisted in the desire of the exclusive possession of the females; and these have acquired weapons to combat each other for this purpose, as the very thick, shieldlike, horny skin on the shoulder of the boar is a defence only against animals of his own species, who strike obliquely upwards, nor are his tusks for other purposes, except to defend himself, as he is not naturally a carnivorous animal. So the horns of the stag are sharp to offend his adversary, but are branched for the purpose of parrying or receiving the thrusts of horns similar to his own, and have therefore been formed for the purpose of combating other stags for the exclusive possession of the females; who are observed, like the ladies in the times of chivalry, to attend the car of the victor.

'The birds which do not carry food to their young, and do not therefore marry, are armed with spurs for the purpose of fighting for the exclusive possession of the females, as cocks and quails. It is certain that these weapons are not provided for their defence against other adversaries, because the females of these species are without this armour. The final cause of this contest among the males seems to be, that the strongest and most active animal should propagate the species, which should thence become improved.

'Another great want consists in the means of procuring food, which has diversified the forms of all species of animals. Thus the nose of the swine has become hard for the purpose of turning up the soil in search of insects and of roots. The trunk of the elephant is an elongation of the nose for the purpose of pulling down the branches of trees for his food, and for taking up water without bending his knees. Beasts of prey have acquired strong jaws or talons. Cattle have acquired a rough tongue and a rough palate to pull off the blades of grass, as cows and sheep.

Some birds have acquired harder beaks to crack nuts, as the parrot. Others have acquired beaks adapted to break the harder seeds, as sparrows. Others, for the softer seeds of flowers, or the buds of trees, as the finches. Other birds have acquired long beaks to penetrate the moister soils in search of insects or roots, as woodcocks; and others, broad ones to filtrate the water of lakes, and to retain aquatic insects, as ducks. All which forms seem to have been gradually produced during many generations by the perpetual endeavour of the creatures to supply the want of food, and to have been delivered to their posterity with constant improvement of them for the purposes required.

'The third great want among animals is that of security, which seems much to have diversified the forms of their bodies and the colour of them; these consist in the means of escaping other animals more powerful than themselves. Hence some animals have acquired wings instead of legs, as the smaller birds, for the purpose of escape. Others, great length of fin or of membrane, as the flying-fish and the bat. Others, great swiftness of foot, as the hare. Others have acquired hard or armed shells, as the tortoise and the *Echinus marinus* [the sea-urchin]. Mr Osbeck, a pupil of Linnæus, mentions the American frog-fish, *Lophius histrio*, which inhabits the large floating islands of seaweed about the Cape of Good Hope, and has fulcra resembling leaves, that the fishes of prey may mistake it for the seaweed which it inhabits.

'The contrivances for the purposes of security extend even to vegetables, as is seen in the wonderful and various means of their concealing or defending their honey from insects, and their seeds from birds. On the other hand, swiftness of wing has been acquired by hawks and swallows to pursue their prey; and a proboscis of admirable structure has been acquired by the bee, the moth, and the humming-bird, for the purpose of plundering the nectaries of flowers. All which seem to have been formed by the original living filament, excited into action by the necessities of the creatures which possess them, and on which their existence depends.'

It is a matter of great interest to note in the foregoing passage that Erasmus Darwin had got hold of one side of the principle which his grandson subsequently elaborated into his theory of 'Sexual Selection'—the principle, namely, that certain structural peculiarities can be acquired, and when acquired may be intensified in the process of inheritance, owing to the fact that only those males possessing the peculiarity have the opportunity of leaving descendants. Erasmus Darwin, however, ascribed to a 'final cause' what Charles Darwin would have regarded as a result. We also see in the above, that Erasmus Darwin clearly recognised the significance and importance of what are now called 'protective resemblances.'* On the other hand, he curiously inverts the case where he speaks of the contrivances by which plants protect or conceal their honey from insects; and no one has done more than his own

* In other passages he gives a much fuller account of protective resemblances among animals, and adduces many instances, as to which he says that though the 'final cause' is easily understood, 'the efficient cause would seem almost beyond conjecture.'

grandson to prove how numerous, varied, and complex are the contrivances by which many plants *attract* insects, for the purpose of having their seeds fertilised.

In another passage, Erasmus Darwin makes special note of the extraordinarily rapid multiplication of living beings, and he recognised that the great majority of the young of each species must of necessity perish before reaching maturity. In this, however, he saw nothing more than a provision of nature to prevent the species, as a species, from suffering extinction. He failed, therefore, just at the point where Charles Darwin succeeded; and he does not appear to have suspected that it is this fact which forms the starting-point of the long series of causes concerned in the origination of new species. No traces, in fact, of the law of 'Natural Selection,' as subsequently set forth by Charles Darwin, can be detected in his utterances upon this subject.

With regard to the general conclusions at which Erasmus Darwin arrived, he concluded that 'all animals have a common origin, namely from a single living filament, and that the difference of their forms and qualities has arisen only from the different irritabilities and sensibilities, or voluntarities, or associabilities, of this original living filament.' Hence he thought it 'not impossible but the great variety of species of animals which now tenant the earth may have had their origin from the mixture of a few natural orders.' Indeed, he goes further than this would imply, since he says in a later passage: 'From thus meditating on the great similarity of the structure of the warm-blooded animals, and at the same time of the great changes they undergo both before and after their nativity; and by considering in how minute a portion of time many of the changes of the animals above described have been produced; would it be too bold to imagine that all warm-blooded animals have arisen from one living filament, which THE GREAT FIRST CAUSE endued with animality, with the power of acquiring new parts, attended with new propensities directed by irritations, sensations, volitions, and associations; and thus possessing the faculty of continuing to improve by its own inherent activity, and of delivering down those improvements by generation to its posterity, world without end?'

It is clear, therefore, that Erasmus Darwin not only taught the doctrine of the origin of species by descent with modification, but he regarded the course of development as an ascending one. This is rendered quite certain by a still later passage, in which he expresses the opinion that, 'from the beginning of the existence of this terraqueous globe, the animals which inhabit it have constantly improved, and are still in a state of progressive improvement.' He adds, 'this idea of the gradual generation of all things seems to have been as familiar to the ancient philosophers as to the modern ones; and to have given rise to the beautiful hieroglyphic figure of the πρωτον ωον, or first great egg, produced by NIGHT—that is, whose origin is involved in obscurity, and animated by ερος, that is, by DIVINE LOVE; from whence proceeded all things which exist.'

Not only did Erasmus Darwin accept the principle of evolution as applied to living beings; but he quotes with approval the idea which Hume had put forth, that the globe itself 'might have been gradually produced from very small beginnings, increasing by the activity of its inherent principles rather than by a sudden evolution of the whole by the Almighty fiat.' Nor did he regard this view as being in the smallest degree antagonistic to the Theistic conception of the universe; for he adds: 'What a magnificent idea of the infinite power of THE GREAT ARCHITECT! THE CAUSE OF CAUSES! PARENT OF PARENTS! ENS ENTIUM!'

THE TRANSMUTATION OF SPECIES.

LAMARCK.

UP to the middle of the present century, naturalists in general had regarded 'species,' both of animals and plants, as immutable entities, founded upon abstract conceptions in the mind of a Creative Being, and necessarily incapable of modification except within the narrowest limits. It was held that each species had been created with a determinate and invariable organisation, by which it was specially adapted to the particular region which it inhabited. On this view, the habits of the species were the necessary result of its organisation, and as the latter was believed not to vary, so it was assumed that the former were also invariable. We have seen that Edward Forbes held this opinion, and that it was to the same doctrine that Cuvier lent the support of his immense influence and his vast learning.

At the present day, it is questionable if there exist any naturalists who regard 'species' as being independent creations, in the sense in which Cuvier and Edward Forbes held that they were. The principle of evolution, on the other hand, is, in one form or another, almost or quite universally admitted. Opinions may and do differ as to the extent to which evolution has operated, and as to the precise method or methods in which the process has been carried out. Admittedly, also, there are still many difficulties remaining unexplained, or only partially explained, by any theory of evolution. All naturalists, however, are now willing to admit that the existing species of animals have been produced by the gradual modification of pre-existing species, and that these, in turn, have been evolved from a still older series of specific forms.

The successful accomplishment of this revolution in the Philosophy of Zoology must be associated with the great name of Charles Darwin. We have seen, however, that the application of the principle of evolution to the solution of the problems of organic life was first systematically attempted by Erasmus Darwin, at the end of last century. In the beginning of the present century, the same principle was applied, more rigidly, and with greater completeness, to the problem of the origin of species by the celebrated French zoologist, De Lamarck, whose life and writings may be briefly glanced at here.

Jean Baptiste Pierre Antoine de Monet, usually known as the Chevalier de Lamarck, was born on the 1st of August 1744, at Bezantin, a small village in Picardy. He was of noble descent, but his father was poor, and being the youngest of a large family he was educated with a view to entering the church. He had, however, an invincible repugnance to a clerical life, and when his father died in

1760, he at once abandoned his studies, and betook himself straightway to the French army, which was at that time in Germany. He was admitted as a volunteer in a regiment of Grenadiers, and the next day distinguished himself so highly on the field of battle, that he was at once made an officer. His military career was, however, shortly thereafter brought to an abrupt close, in consequence of an injury inflicted upon him in sport by one of his companions, which necessitated the performance of a severe surgical operation, and left him permanently unfitted for the life of a soldier. He was therefore compelled to earn his own living as best he could, and to begin with he became a banker's clerk, a position in which he remained a long time. For four years he studied medicine, but he ultimately abandoned this, and devoted the whole of his spare time to scientific studies, his favourite subjects at first being meteorology and botany.

To the study of botany in particular Lamarck applied himself with the greatest diligence; and in 1778 he published his first book, the 'Flore Française.' This was a sort of descriptive catalogue of all the known species of plants in France, and its special object was to allow of the ready identification of any unknown plant. The method of arrangement adopted in this work is what has been called the 'dichotomous' or 'binary' arrangement, a method which has the merit of great simplicity, though in itself essentially artificial. Lamarck was perfectly acquainted with the natural system of classification of plants, which had been introduced into botany by the Jussieus; and he fully recognised the merits of this system as a philosophical arrangement. He recognised, however, that the very merits of the natural classification render it difficult of use by beginners. A natural arrangement cannot be employed without difficulty for the mere purpose of identifying an unknown species. On the other hand, an artificial system, like that introduced into botany by Linnæus, can be used with the utmost ease in the identification of species, though it has the philosophical demerit of associating together forms which have no real relationships to one another. The method employed by Lamarck in the 'Flore Française' was a sort of combination of the natural and artificial methods of classification. The 'dichotomous' or analytical method consists, in fact, in a classification of natural objects by means of positive and negative characters, the characters selected being always obvious and easily recognised. In the identification of a species, therefore, the choice is always restricted to one of two opposite characters, and the method proceeds by constantly dividing and subdividing by two. The dichotomous method of classification, as adopted by Lamarck in botany, has been used also in natural history. When one has to deal with very large groups of nearly allied species, such, for example, as the different groups which constitute the order of the Beetles, or in the case of the different species of a single large genus, the dichotomous method is sometimes extremely useful. It is, however, distinctly an artificial method, and labours under the inherent weakness that a group defined by a negative character only *must* be artificial in principle, and is hardly likely to prove in actual practice to be natural even by accident.

At the time when Lamarck published his 'Flore Française,' botany, owing largely to the influence of Rousseau, was a favourite subject of study in France; and the work was very favourably received. Its publication led to his obtaining the friendship and patronage of Buffon, and also to his being appointed to a subordinate place in the botanical department of the Academy of Sciences. Buffon likewise sent him, as tutor to his son, on a tour through Europe, and obtained for him at the same time a sort of official commission to visit foreign botanical gardens and museums, whereby he much increased his knowledge of plants and animals. Lamarck now continued to prosecute diligently his studies on plants, and published some large and excellent botanical treatises; but unfortunately he remained poor, and without any remunerative or permanent employment. Ultimately he was appointed curator of the botanical collections of the 'Cabinet du Roi.' In 1793, however, even this ill-paid post was suppressed by the National Assembly, when the 'Cabinet du Roi' became converted into the Museum of Natural History and the 'Jardin des Plantes.' In this new institution Lamarck was called upon to take the professorship of the Vermes and Insecta of Linnæus, or, as we should now say, of the Invertebrate Animals.

At this time, the Invertebrate animals were exceedingly ill understood, if we except the *Insecta*, being, in fact, very much in the condition in which they had been left by Linnæus. Besides, Lamarck, who was now fifty years old, had not previously studied zoology closely, and was very imperfectly acquainted with any of the groups of the Invertebrates, except the Molluscs, the shell-fish having always been one of his favourite subjects of study. It was characteristic of the ardent and enthusiastic temperament of the man, and of the thoroughness with which he threw himself into his work, that starting in this way, in advanced life, and with a subject almost new to him, he should have been able finally to give to the world that gigantic and classical exposition of his special department known as the 'Histoire Naturelle des Animaux sans Vertèbres.' The first edition of this great and famous work was published between the years 1815-1822, in seven octavo volumes, of which five were written wholly by Lamarck himself. Owing to the failure of his sight, he was assisted in the part relating to the insects by Latreille; and when his sight wholly gave way, the remainder of the work was written by Valenciennes, or drawn up by one of his daughters from his previously written notes and papers. The second edition of this work, edited by Deshayes and Henri Milne-Edwards, is the one now generally used. It is in eleven volumes octavo, and was published between 1835 and 1845. It is not necessary here to enter into any detailed analysis of this wonderful work. It is still indispensable to working zoologists, in almost all the groups of Invertebrate animals; and the classification adopted in it is in many respects a great improvement upon any that had preceded it. His primary division of the animal kingdom into 'Apathetic,' 'Sentient,' and 'Intelligent' animals, even if it expresses an underlying truth, is one practically inapplicable. It is, however, to Lamarck that we owe the exceedingly useful general name of 'Invertebrate Animals' for the entire series of animals below the Vertebrata.

Lamarck's later life was an unfortunate one, and it was greatly to the discredit of his countrymen, usually so prompt to recognise and reward genius, that it should have been so. He was attacked by a disease of the eyes, which at first led only to an impairment of vision, but which terminated in total blindness. His limited savings had been lost in some unlucky investment, and he was thus deprived not only of his employment but also of the means of life. His closing years were therefore passed in extreme poverty and in helplessness, though the latter affliction was to some extent alleviated by the devoted affection of one of his daughters. He died on the 18th of December 1829, in the eighty-fifth year of his age.

The name of Lamarck can never be forgotten in the history of zoology; and his great work on the Invertebrate Animals would alone have been sufficient to secure this. He also left a permanent mark in certain special groups of Invertebrates, and particularly as regards the *Mollusca;* his labours as to living and fossil shell-fish having been of great extent, and having produced results of great value. Lamarck, however, was very much more than a mere observer, describer, or classifier of animals. He possessed a singularly original mind, prone to generalisation, and bold to rashness in its conceptions. In all the subjects which he touched, he showed this tendency; and it is no matter for surprise to find him the author of such books as the 'Système analytique des Connaissances positives de l'Homme' and the 'Philosophie Zoologique,' to say nothing of his writings on the Theory of Chemistry or on Hydrogeology. The most famous of his philosophical or theoretical treatises is his 'Philosophie Zoologique,' published in 1809, in two volumes octavo; and the fame of this rests in large part upon the fact that it was here that he first laid down what may be regarded as the earliest definite theory of evolution, as applied to living beings.

The only parts of Lamarck's theoretical views with which we are specially concerned here, are those relating to the nature of 'species,' and as to the way in which species originated. Up to this time, naturalists in general believed that though 'species' might be capable of variation, such variations were only possible within certain definite limits. On this view, a species might oscillate backwards and forwards on both sides of a central line, but it would sooner or later return to the position of equilibrium represented by the *type-form* of the species. This view was the one held by the illustrious Cuvier, and by many naturalists long after the time of Lamarck. Many, indeed, believed that 'species' of animals and plants were *special creations*, and as such necessarily incapable of transmutation. This, for example, was, as has been seen, the view held by Edward Forbes. A few observers (such as Dr Erasmus Darwin, Isid. Geoffroy St Hilaire, and Goethe) had begun to doubt the stability or permanence of species as early as the end of the eighteenth century. Lamarck not only reached this belief independently, but clearly formulated it, first in 1801, then in the 'Philosophie Zoologique' in 1809, and finally in the introduction to the 'Animaux sans Vertèbres' in 1815.

Lamarck was led to his views as to the transmutability of species by his study of *varieties*, the ever-recurring stumbling-block of systematic zoologists. The general ideas which are in the minds of naturalists when they speak of 'species' are expressed by the definition, that a 'species' of animals consists of *an assemblage of individuals, all resembling each other, and producing their like by generation.* Thus, all wolves * resemble each other and produce fertile young; hence wolves constitute a single 'species,' the *Canis lupus* of Linnæus. Similarly, the individuals of the lion, tiger, brown bear, and so on, resemble each other and produce fertile young, thus constituting so many distinct 'species' of animals. There is, however, the obvious difficulty that the individuals of any species do not *altogether* resemble each other. They resemble each other in generals, but not in particulars. All wolves are alike, but they are not absolutely alike. Besides, in many cases the differences which exist in the individuals which compose any given species may be very considerable; and so far as our observation goes, they may be permanent differences. This is most conspicuously the case among our domesticated animals, such as the dog or pig. Such species are known by every one to contain certain groups of individuals which differ extraordinarily from each other; and the differences distinctive of these groups are to all appearance as persistent as the differences which separate distinct species. Thus, to take the case of dogs, the mastiff, greyhound, bulldog, terrier, and so on represent such groups of differing individuals, which in common language are known as different 'breeds' or 'races' of dogs. Such breeds are in some cases known to have existed without notable

* The European wolves, of course, are meant.

change from the very earliest historical times onwards. Though, for readily intelligible reasons, seen in a more marked form in domestic animals than in wild ones, such groups of differing individuals are seen in a vast number of species of animals from the lowest to the highest. Most 'species' of animals, therefore, include one or more 'varieties;' and this is a phenomenon quite as well seen in the vegetable kingdom as amongst animals.

Now, the question arises—at what point do the differences which distinguish a 'variety' from a 'species' become so pronounced, that we conclude that we have a fresh *species* to deal with, and not a mere *variety* of an old species? This is a point upon which naturalists have not as yet succeeded in laying down any fixed rules. From the nature of the problem, it is extraordinarily difficult to detect any underlying principle to guide us in practice in deciding between species and varieties. Hence, in what are called 'variable' genera (such, for example, as the genus *Rubus*, including the brambles, or the genus *Salix*, containing the willows), observers have never been able to agree as to the precise number of species which exist, even in a small country like Britain —since what one observer sets down as two species, another equally competent authority will regard as being only a species and its variety. The test usually adopted by naturalists in distinguishing between varieties and species is what has been called the 'physiological test.' That is to say, when the individuals of a given assemblage of animals or plants are fertile, and are capable of giving rise to fertile offspring, then they are usually regarded as constituting a single *species*, however greatly they may differ among themselves in structure or appearance. 'Varieties,' therefore, are supposed to be always capable of interbreeding with the type-form of the species. On the other hand, if two groups of animals or plants, otherwise nearly resembling each other, are found to be incapable of producing fertile offspring by intercrossing, then they are regarded as constituting *two* distinct species. The physiological test, however, is wholly inapplicable to extinct organisms, where the same difficulties as to the distinction between species and varieties exist as among living forms. It also cannot be applied without some serious reservations even as regards living beings, since some organisms which are otherwise clearly recognisable as distinct species, are undoubtedly capable of producing fertile offspring by interbreeding. This is seen in the case of various plants, and in a few instances among animals.

Lamarck was profoundly impressed with the difficulty of separating species from varieties, and also with the very wide range of the variability shown by many animals and plants. To this point he recurs again and again. 'In proportion,' says he, 'as we gather together the productions of nature—in proportion as our collections become more and more extensive, in the same proportion do we see blank after blank filled up and our lines of separation effaced. We find ourselves reduced to an arbitrary determination, which at one time compels us to seize on the most minute varietal differences as characters of what we call a *species*, while at another time it forces us to set down those slightly differing individuals of which other observers make a species, as really nothing more than a *variety*.' Every working naturalist will heartily concur in the truth of these remarks.

Lamarck, therefore, though willing to accept the ordinary definition of a 'species' as 'a collection of similar individuals which were produced by individuals like themselves,' rejected the idea of the constancy of species. He maintained that those characters of a species which we call 'specific' were liable to variation, and he supposed that this variation was indefinite. Hence, he maintained that the constancy of species is not absolute, but only relative to the circumstances in which all the individuals of the species are placed.

Species, then, according to Lamarck, represent groups of individuals which are only stable so long as their environment remains essentially unchanged. That species should appear to us to be permanent, he explained upon the ground that our observations had only extended over a few thousand years, and that this period had not been long enough to allow of the transformation of any one species into any other, especially as terrestrial changes have been quite slight and unimportant during the whole period embraced by human observation.

'A multitude of facts,' says Lamarck, 'teach us that in proportion as the individuals of our "species" change their locality, their climate, their manner of living, or their habits, in the same proportion they become subject to influences which bit by bit change the consistence and proportions of their parts, their form, their faculties, even their organisation, in such manner that, given sufficient time, everything in them participates in the mutations to which they are exposed.

'In the same climate, the first result of a wide difference of habitat or environment is to give rise to simple variation of the individuals affected by these differences. But in the progress of time the prolonged difference of habitat in individuals which go on living and reproducing themselves under similar conditions, gives rise in them to differences which become to some extent essential to their existence. Hence, in the course of many succeeding generations, individuals which belonged, to begin with, to one species, find themselves ultimately transformed into another, new and distinct species.

'Suppose, for example, that the seeds of a grass, or of any other plant natural to a humid meadow, should by any chance be suddenly transported to the slope of a neighbouring hill, where the soil, although more elevated, was still sufficiently moist to allow of the plant maintaining its existence. Suppose, further, that after having lived there, and propagated itself there for a number of times, it should by degrees reach the dry and almost arid soil of a mountainous ridge. If the plant should succeed in finding a subsistence in its new habitat, and should perpetuate itself during a series of generations, it would ultimately become to such an extent modified, that the botanist who might find it would make out of it a new *species*.

'The same thing happens in the case of animals which have been forced to change their climate, their manner of living, and their habits; but in their case the causes which I shall subsequently cite demand a still longer time than among plants, before their influence can produce noteworthy changes in individuals.'

Holding the above opinions as to the stability of species, Lamarck necessarily wholly abandoned the view that the animals and plants now in existence had been produced *de novo*, just as and where we now find them. He, of course, similarly rejected the view that the animals and plants of each successive geological period had been created specially, *en masse*, for that period. He admitted partial and local catastrophes; he would hear nothing of universal catastrophes. On the contrary, his views led irresistibly to the belief that the animals and plants of each successive period of the earth's history, including the present period, had been produced by variation from animals and plants previously in existence.

Lamarck was not, however, clear as to the fact that successive races of animals and plants had during the progress of the ages become extinguished, and had been replaced by other different assemblages of living beings. He treats of this subject in a well-known chapter of the 'Philosophie Zoologique,' entitled 'Des espèces dites perdus.' Upon the whole he is inclined to doubt if the means taken by nature to conserve species and races have been so inefficient and imperfect that entire groups could have become extinct. He admits that many fossil animals are certainly distinct from any known living species. Against this, he thinks, may be set off the fact that our knowledge of living animals is still very imperfect. In the case of the *Palæotheria*, and other large Mammals described by Cuvier from the Tertiary rocks of the Paris basin, he admits that it is impossible for living representatives to exist, and yet to have escaped observation. As regards these he is willing to admit extinction; but he thinks that they have been exterminated by man.

On the other hand, in the case of marine animals, such as shell-fish, where human agency certainly cannot have caused extinction, he has a different theory. Such species, he says, should not be regarded as really extinct, but should rather be considered as belonging to existing species—in the sense, namely, that the living types are only the modified descendants of the fossil forms. The real fact which, in his opinion, ought to astonish us is to find among fossils *any* species which are identical with living forms. 'This fact, which our collections place beyond doubt, ought to lead us to the belief that the fossil remains of animals which are still represented in the living state, are those of which the antiquity is least. The species to which such forms belong have, doubtless, not yet had time to permit of their having varied to any great extent.' Lamarck, therefore, had got hold of the principle which Lyell afterwards employed with such effect in his well-known classification of the Tertiary rocks.

Not only did Lamarck hold that the existing species of animals had been produced by the gradual modification of pre-existing species; but he thought that the course of modification had been, on the whole, a progressive and not a retrograde one. He believed that the simpler forms of life had been produced first, and the more complex ones later—Man being the last evolved and the highest of all animals. In order to account, however, for the existence at the present day of *any* simple or degraded types of animal life, Lamarck thought it necessary to assume that such simple forms were always being produced afresh by spontaneous generation. Otherwise he thought that, in the course of progressive modification, all the lowest forms of living beings would long ere this have been 'improved out of existence' altogether. There is an interesting and curious resemblance between the speculations of the French philosopher on this and kindred subjects and the notions which have been put forth by some writers even in the last score of years.

The subject of the *agencies* which are concerned in the production of modifications in species was dealt with by Lamarck in a most interesting chapter of the 'Philosophie Zoologique,' 'on the influence of circumstances on the actions and habitudes of animals, and of these actions and habitudes as causes which modify the organisation and structure of animals.' The title of this chapter will of itself show that he regarded the general organisation of an animal as, in the long run, the outcome and expression of its actions and its habits; and he regarded these latter as the direct result of the animal's environment. Hence, in investigating the causes which have led to the production of 'species,' Lamarck began with a consideration of the action of *external conditions*.

It is an old and a natural idea that the organs of animals were given to them for the purpose of adapting them to the conditions under which they may happen to be placed. Thus it is natural to the human mind to think that the polar bear is white because it was created to live among the snows of the Arctic regions; or that the lion is tawny in colour because it is meant to live in sandy deserts; or that the long neck of the giraffe is a pre-designed structure adapting it for feeding upon the foliage of trees. Lamarck regarded this view of the subject as a putting of the cart before the horse. He thought that the adaptation between an animal and its environment resulted from the fact that the animal had, so to speak, been compelled to suit its structure to its surroundings. He thought that it was the external conditions which had gradually evoked the corresponding organ or structure. In other words, he thought it was the necessity for action which had produced the corresponding parts.

The contrast between the older view on this subject and that held by Lamarck may be shown by a single example. The giraffe will answer the purpose very well. The giraffe,* as is well known, lives in a region where droughts are of common occurrence, and where therefore the herbage is very liable to become burnt up and destroyed. Being a large animal, it requires considerable quantities of vegetable food; and being gregarious in its habits, it is clear that it could not exist except in a country where plant-life flourished luxuriantly. Owing, however, to the length of its neck, it is enabled to survive the long African droughts, since it can browse upon lofty shrubs and trees, which are not much affected even by a prolonged want of rain.

Now, the older view of the matter would regard the long neck of the giraffe as being a structure aboriginally possessed by the species, and, in fact, specially given to it for the purpose of adapting it for the conditions of its life in its African home. On Lamarck's view of the

* The giraffe is one of the instances given by Lamarck of animals whose peculiarities have been produced by their environment, but his remarks on this particular case are here amplified.

subject, on the other hand, it might be supposed that the giraffe did not in the first instance possess the extraordinarily elongated neck by which it is now characterised; but that it resembled the ordinary Ruminants, to which it is otherwise allied, in having a cervical region of the normal length. It might further be supposed that, to begin with, the climate of Africa was moister than it now is, and that the regions inhabited by the giraffe were not liable, as they now are, to prolonged periodic droughts. If we next assume that the climate of Africa underwent gradually a change, in consequence of which it became drier, and droughts became more frequent, it is clear that the existence of the giraffe would be rendered difficult or precarious, because its short neck would not allow it to reach the higher shrubs, and the ordinary ground herbage might be destroyed for many months together.

According, however, to the conceptions of Lamarck, the increasing frequency and severity of the droughts would give rise to a gradual elongation of the neck of the giraffe *in successive generations*, which would enable the animal to dispense with the periodically destroyed herbage, and to feed habitually on the foliage of trees and shrubs. This gradual lengthening of the neck would not, it need hardly be added, be produced in one individual, but would be the result of the progressive elongation of the cervical vertebræ in a long series of generations. Hence, on Lamarck's view, the long neck of the giraffe would have to be regarded as the direct result of the surroundings of the animal, and not as a pre-designed structure intended to meet foreseen conditions.

We may pause for a moment here to ask the bearing of the above explanation of the long neck of the giraffe on the ordinary doctrine of design as displayed in nature. In the coarse and rude teleology of the first half of the present century, and of preceding periods, the long neck of the giraffe would have been adduced as one striking proof out of many, that the Creator had produced each species of animals as we now find it, and had specially fashioned each to meet the conditions present in the area within which it had been created. To a teleology of this sort Lamarck's view is necessarily and absolutely fatal. At the same time it is to be remembered that Lamarck's theory is entirely compatible with the belief in the existence of a Creator and of design in nature. It is possible, namely, to believe that the power enjoyed by the neck of the giraffe of lengthening under a given set of conditions formed an integral part of the original design of the animal, or of the species from which it was evolved. It may even be held that the design which embraces the power of indefinite adaptation to varying conditions is of an infinitely higher order than the design which merely adapts an animal for one set of conditions, it being beforehand certain that those conditions must ultimately give place to a different set.

In the same way, a much higher conception underlies the theory of creation by evolution than is involved in the older view of the separate creation of each species. Lamarck has left us in no doubt as to what his own views were on this point. Nor is there the smallest reason to doubt the genuineness of his words, or to think that his remarks—like some of those made by Buffon—were merely intended to disarm the ecclesiastical prejudices of his day.

'Without doubt,' says he, 'nothing exists save by the will of the sublime Author of all things. But are we able to lay down laws for Him in the execution of his will, or to fix the methods in which He has carried out his purposes? Why should not His infinite power have been able to create *an order of things*, which has successively given rise to everything which we see, as well as to everything which exists, but of the existence of which we are ignorant?'

Strictly speaking, according to Lamarck, *circumstances* have in themselves no power of directly modifying the organisation of an animal. Changed circumstances, however, give rise to changes in the *needs* ('besoins') of animals; changed needs imply and necessitate changed *actions* on the part of animals; and if the changes of circumstances become permanently established, so also do the changes of habitudes thence resulting. But change of habit implies corresponding change on the part of the animal as to the organs which it most largely employs. If an animal has enjoyed a particular habit of life, it has necessarily used the organs which conform to that habit and are in agreement with it. If, on the other hand, it changes its habit of life, it must use more sparingly, or cease to use, the organs which it formerly exercised, and it must call into play another set of organs.

In the case of plants, which have no *actions* (in the sense in which Lamarck used this term), and which then cannot be properly said to have any *habitudes*, changed conditions similarly produce changes of structure, some organs becoming thereby more developed, while others diminish or altogether disappear. In this case, however, changed circumstances operate by inducing changes in the nutrition of the plant, 'in its processes of absorption and transpiration, in the quantity of heat, light, air, and moisture which it habitually receives, finally in the predominance which may be established in some of its vital movements over others.' Lamarck brings forward many instances of the changes produced in plants by changed circumstances, and expresses the opinion that in plants such changes are more rapidly brought about than in animals; since the causes which affect animals operate very slowly, and are consequently difficult of appreciation and recognition by us.

With regard to the nature of the circumstances which are mainly operative in producing changes in animals, Lamarck assigns the first place to the character of the medium in which it lives—whether it be terrestrial, aërial, or aquatic in its habits. He also attaches great importance to the differences of temperature, moisture, and the like in different regions, and thinks that such differences are a great cause of variation in animals and plants. Every one, he says, admits these great climatic differences; but 'what is not sufficiently recognised, or is even altogether denied, is that each *place* itself is liable in the course of time to changes of its climatic conditions; these changes being effected with such slowness, in relation to our lifetime, that we attribute to the existing conditions a perfect *stability*.' To this he elsewhere adds that 'we may be sure that this appearance of stability in natural things will always be taken, by the generality of mankind, as *real*, because in general our judgment is only relative to ourselves.' Our notion of the permanence of 'species' is based, therefore, upon our idea of the stability of the conditions existing in each area; since species remain unchanged so long as their environment remains the same, and the latter changes so slowly as to elude our powers of observation.

Lamarck's doctrine that useful structures in animals are really the result of the actions of surrounding conditions, has been interpreted as implying that such structures could be developed by repeated acts of *volition* on the part of an animal. This, however, is a travesty of his actual views,* and arises from confounding the *requirements* ('besoins') of an animal, and the actions thence resulting, with its *wishes*. As a good deal of misconception has prevailed upon this point, it may be well to quote one of the passages in which Lamarck expresses his views in a concrete form:

'I propose to show,' he remarks, 'that the continued use of an organ, together with the efforts made by the animal to adapt the organ for the purposes which its surroundings render necessary, will strengthen, develop, and enlarge that organ, or will create new organs which are capable of discharging the functions which have become needful.

'The bird which necessity conducts to the water for the purpose of obtaining its food, separates its toes when it wishes to strike the water or move upon its surface.

* This has been lately insisted upon by Professor P. Martin Duncan, in the anniversary address to the Linnean Society for 1884, in which an excellent summary of Lamarck's philosophico-zoological views is given.

The skin which unites the toes at their bases acquires, in consequence of the ceaselessly repeated separation of the toes, the power of extension. Thus, in the course of time, are formed the extended webs which unite the toes of ducks, geese, and the like. Similar efforts to swim, that is to say, to progress in a fluid element by striking the water, have in like manner given rise to the extension of the skin between the toes in frogs, turtles, otters, beavers, and such like animals.

'On the contrary, a bird which is habituated by its mode of life to perch on trees, and which is descended from individuals all of which have had a similar habit, necessarily possesses toes more elongated than, and differently constructed to, those of the aquatic animals just alluded to. Its claws, in course of time, become lengthened, pointed, and curved, so that it can grasp the branches of trees on which it so often reposes.

'In the same way, a bird frequenting the shore, which does not wish to swim, but which nevertheless is obliged to approach the margin of the water for the purpose of catching its prey, is continually exposed to plunging into the mud. Such a bird, then, being desirous to save its body from becoming wet, makes every effort to extend and lengthen its legs. From the long-continued habit of extending the legs, contracted by this bird and by all of its species, it results that the individuals of the species are now found elevated, as it were, upon stilts, having gradually obtained long naked legs, destitute of feathers up to the thighs or even higher.

'Again, the same bird, wishing to fish without wetting its body, is obliged to make continual efforts to lengthen its neck. Now, the effect of such habitual efforts in such a bird, and in all of its species, ought to be that of producing in time a marked elongation of the neck; and, as a matter of fact, such lengthening of the neck is actually found in all shore-frequenting birds.'

It will be evident from the above, as from the entire tenor of his arguments upon this subject, that Lamarck does not mean that a mere act of volition, apart from any positive action, could effect any change in the structure of an animal. He assuredly did not mean, for example, to imply that the long legs of wading birds were the result of the fact that repeated generations of these birds had gone on wishing that their legs were a little longer. All that his theory would imply in such a case is, that when a change in the previous condition of a region had driven certain formerly terrestrial birds into the habit of wading in streams, pools, or the sea, for the purpose of obtaining food, the necessity for keeping their bodies dry had ultimately brought about a lengthening of the bones of the leg, and a disappearance of part of their covering of feathers. In other words, Lamarck really means that the constant efforts made by any and every animal to bring itself into accord with its surroundings inevitably leads in course of time to corresponding changes of structure.

This point may be illustrated by an imaginary case. As things stand at present, man has no necessity for using his foot in prehension, his hand supplying all his wants in this direction; and his great toe is therefore practically useless as an organ of prehension. It is, however, a well-known fact that men who leave the foot in its natural condition, uncramped by artificial coverings, have some power of using the great toe as a hinder thumb. This is exemplified in many sailors, and in various savage tribes; and in such cases the prehensile great toe is of much assistance in climbing. It is also well-known that in individuals who have never possessed, or who have lost the use of the hands, the great toe can be trained by constant use to act as a thumb, and the foot can thus be converted into a more or less efficient hand. Suppose, however, that in some particular region, circumstances should arise which should render it advantageous or necessary for men to quit the ground, and to take up their abode permanently in trees. Such circumstances can be quite easily imagined. In such a case, it would be of enormous advantage to the individual that he should be able to use his great toe in grasping, so as to help him in climbing. We may assume, therefore, that each individual would practise himself, consciously or unconsciously, in using his feet as prehensile organs. We may also assume that each individual would in the course of time succeed in these prolonged attempts, and would ultimately acquire a more or less complete power of 'opposing' the great toe to the other digits, and so of using the foot to grasp branches in climbing. Supposing, however, that the conditions which drove men to reside in trees became permanent, we cannot doubt that in each successive generation this power of using the foot in grasping would improve. We may rest assured that repeated efforts to perform a particular function with a particular organ must result in an increased ability on the part of the organ to do the work demanded from it. It is, moreover, certain that *pari passu* with the change of *function* of the great toe there would take place a modification of its *structure*. It would gradually come to be placed more and more at an open angle to the other digits of the foot; its muscles would become more and more developed, giving it increased power and range of movement; an 'opponent' muscle might be developed; and, in all probability, the structure of the ankle would become so far modified that the sole of the foot could readily be turned inwards.*

The above case is, of course, a purely imaginary one; but there is nothing in it which would be inconsistent with universally admitted biological laws. It is, however, also strictly consistent with the Lamarckian theory; since it affords an instance in which repeated efforts on the part of the individuals of successive generations to use a particular organ in a particular way would result in the gradual change of that organ, both as regards function and structure, so as to suit it to the new requirements (*besoins*) of the individual.

As to the question *how* changes in the surrounding conditions should give rise to corresponding changes in the structures and organs of animals, Lamarck relied principally upon the effects of use and disuse. No doctrine in physiological science is better established than that which teaches us that the habitual use of an organ leads to a corresponding growth on the part of the same. If we employ a particular muscle much, it

* This power of inverting the sole of the foot is naturally possessed by the young of the human species, as is also the power of freely moving the great toe, both being more or less extensively lost in later life.

will increase both in size and weight, and will at the same time become more fully competent to discharge its particular work. On the other hand, disuse of a muscle leads, in the first place, to a decreased ability on the part of the organ to perform its proper function. Thus, the muscles of the left hand cannot in ordinary people be used as efficiently as those of the right hand, because they are less often used. If the disuse be prolonged and habitual, the organ will become diminished in size, and the ability to discharge its function may be wholly lost. Thus, man possesses the three muscles which are attached to the external ear, and which enable many of the quadrupeds to move their ears freely; but in him they are extremely reduced in size, and the power of employing them has, from disuse, become almost, or entirely, lost. Finally, if the disuse of an organ be complete, as when external conditions no longer demand its employment, it may become a mere rudiment, absolutely destitute of function. This is the case, for example, with the eyes of certain animals which spend their existence living underground and in total darkness. Lamarck fully recognised the important results which flow from the use or disuse of organs; and it was, therefore, to the effects of *habit* that he chiefly ascribed the progressive modifications which he believed to have affected the structure of all living beings. He believed that a change in the surrounding conditions would necessarily compel an animal to modify its former habits of life; and that a change in its habits would necessarily entail an increased use of certain organs and a decreased use of others. The organs upon which new and increased demands were made would undergo a proportionate development in size, and would simultaneously become better fitted for their work. At the same time, they might become modified in *form*, and thus might become better suited for the purposes for which they were required. On the other hand, the organs which were called into action less frequently than they had previously been, would gradually diminish in size, and might ultimately disappear. By a continuance of this process through many successive generations, the whole organisation of an animal might become profoundly modified. This was therefore the process by which, in the main, Lamarck believed that 'species' had been evolved out of other pre-existing forms of life.

THE DOCTRINE OF PROGRESSIVE DEVELOPMENT.

THE 'VESTIGES OF CREATION.'

BEFORE proceeding to give a sketch of the theory of 'Natural Selection,' as propounded by Charles Darwin, a brief consideration may be fitly given to the celebrated work entitled the 'Vestiges of Creation.' This well-known treatise was first published in 1844, and the tenth edition appeared in 1853, a sufficient proof of the wide popularity which it enjoyed. Its authorship was never acknowledged during the lifetime of its writer, and it was ascribed to various distinguished persons, with the most varied qualifications for the writing of such a work. Amongst others, Robert Chambers was selected as the author of the 'Vestiges,' and in this instance, as now openly acknowledged, the surmise was correct.* As, however, the work was an anonymous one, the personality of the writer may well

---

* A twelfth edition of the 'Vestiges' was published in 1884 (W. & R. Chambers), containing a preface by Mr Alexander Ireland, in which the facts relating to the authorship are given in detail.

be left out of account in the few remarks which will here be made as to its general scope and teaching.

The 'Vestiges' not only dealt with the origin of the various forms under which vitality has been manifested in the past, or is exhibited at the present day; but it also took up the much wider and much more inscrutable problem of the origin of life itself. It was therefore more than simply an attempt to explain the origin of 'species.' The work is remarkable for the ability which it displays in the handling of general principles, for the closeness of its reasoning, for the clearness with which fallacies are detected and exposed, for the lucidity of its style, and for the wealth of its suggestions. It is, nevertheless, inadequate for the purpose which the writer proposed to himself as the object of his labours. It is, namely, unmistakably, the work of a writer who had mastered the general facts and principles of zoology and the kindred sciences, but who at the same time was without that minute knowledge of biological science which can be obtained in no other way save by long-continued and rigidly conducted first-hand investigation into the phenomena presented by living beings. No fact, indeed, stands out more clearly in the whole history of science than the insufficiency of a merely general knowledge for the establishment of generalisations of any kind. Only the worker whose mind is stored with the minutest details of his subject can safely enter upon the task of generalisation.

Having premised so much, we may briefly sketch the conclusions which are set forth in the 'Vestiges.' After a general review of the relations of the earth to the solar system, and of other astronomical considerations which bear upon geology, an outline is given of the known facts as to the life-history of the earth, and as to the general succession and progression of organic types in past time. With regard to the purely geological history of the earth, the writer of the 'Vestiges' concludes, with Lyell, that 'there is nothing in the whole series of operations displayed in inorganic geology, which may not be accounted for by the agency of the ordinary forces of nature.'

On the other hand, the past history of the earth is not a mere record of physical changes. 'Mixed up with the geognostic changes, and apparently as a final object connected with the formation of the globe itself, there is another set of phenomena presented in the course of our history—the coming into existence, namely, of a long suite of living things, vegetable and animal, terminating in the families which we still see occupying the surface. The question arises—In what manner has this set of phenomena originated? Can we touch at and rest for a moment on the possibility of plants and animals having likewise been produced in a natural way; thus assigning immediate causes of but one character for everything revealed to our sensual observation; or are we at once to reject this idea, and remain content, either to suppose that creative power here acted in a different way, or to believe, unexaminingly, that the inquiry is one beyond our powers?'

In answering this question, the writer of the 'Vestiges' decides unhesitatingly, as every naturalist would at the present day decide, that we cannot consistently accept natural causes as sufficiently explaining the phenomena of the inorganic world, and at the same time invoke supernatural causes to explain the phenomena presented by living beings.

'It is surely,' he remarks, 'very unlikely, à priori, that in two classes of phenomena, to all appearance perfectly co-ordinate, and for certain intimately connected, there should have been *two totally distinct modes of the exercise of the divine power*. Were such the case, it would form a most extraordinary, and what to philosophic consideration ought to be a most startling, exception from that which we otherwise observe of the character of the divine procedure in the universe. Further, let us consider the comparative character of the two classes of phenomena, for comparison may of course be legitimate until the natural system is admitted. The absurdities into which we should thus be led must strike every reflecting mind. The Eternal Sovereign arranges a solar or an astral system, by dispositions imparted primordially to matter; he causes, by the same majestic means, vast oceans to form and continents to rise, and all of the grand meteoric agencies to proceed in ceaseless alternation, so as to fit the earth as a residence for organic beings. But when, in the course of these operations, sea-weed and corals are to be for the first time placed in those oceans, a change in his plan of administration is required. It is not easy to say what is presumed to be the mode of his operations. The ignorant believe the very hand of the Deity to be at work. Amongst the learned, we hear of "creative fiats," "interferences," "interpositions of the creative energy;" all of them very obscure phrases, apparently not susceptible of a scientific explanation, but all tending simply to this—that the work was done in a marvellous way, and not in the way of nature. Let the contrast between the two propositions be well marked. According to the first, all is done by the continuous energy of the divine will—a power which has no regard to great or small: according to the second, there is a procedure strictly resembling that of a human being in the management of his affairs. And not only on this one occasion, but all along the stretch of geological time, this special attention is needed whenever a new family of organisms is to be introduced; a new fiat for fishes, another for reptiles, a third for birds; nay, taking up the present views of geologists as to species, such an event as the commencement of a certain cephalopod, one with a few new nodulosities and corrugations upon its shell, would, on this theory, require the particular care of that same Almighty who willed at once the whole means by which INFINITY was replenished with its worlds.'

It will be seen from the above characteristic quotation that the theory of the 'special creation' of the different species of animals and plants was handled by the writer of the 'Vestiges' in the most uncompromising manner. It is difficult for us, living at a time when naturalists have almost universally abandoned the doctrine of independent creation and of the fixity of species, to appreciate the excitement, the alarm, and the indignation produced by such outspoken deliverances as the above. It needs to be remembered, however, that at the time of the appearance of the 'Vestiges' all the leading British naturalists still adhered tenaciously to the idea that 'species' had been separately produced, and were immutable. We can therefore imagine the consternation of English naturalists generally at the publication, by an unknown writer, of an attack so direct and so unanswerable upon this so cherished doctrine.

Of course, the writer of the 'Vestiges' was by many held up to public obloquy as an obvious materialist, or even as an atheist. It is, however, quite impossible for any one to read the 'Vestiges' with an unbiassed mind, and not to recognise that the writer had discharged his self-appointed task in a spirit at once genuinely reverential and in the truest sense religious. On this point, it is worth while to quote his own remarks. It has been urged, he says, that 'to presume a creation of living beings as a series of natural events, is equivalent to superseding the whole doctrine of the divine authorship of organic nature. With such a notion infesting the mind, it must of course be almost hopeless that the question should be candidly entertained. There can, in reality, be no reason adduced for holding this as necessarily following from the idea of organic creation in the manner of law, or by a natural method, any more than from a similar view of inorganic creation. The whole aim of science from the beginning has been to ascertain law; one set of phenomena after another has been brought under this conception, without our ever feeling that God was less the adorable Creator of his own world. It seems strange that a stand should appear necessary at this particular point in the march of science. Perhaps if our ordinary ideas respecting natural law were more just, the difficulty might be lessened. It cannot be sufficiently impressed that the whole idea relates only to the *mode* in which the Deity has been pleased to manifest his power in the external world. It leaves the absolute fact of his authorship of and supremacy over nature precisely where it was; only telling us that, instead of dealing with the natural world as a human being traffics with his own affairs, adjusting each circumstance to a relation with other circumstances as they emerge, in the mode befitting his finite capacity, the Creator has originally conceived, and since sustained, arrangements fitted to serve in a general sufficiency for all contingencies; himself, of course, necessarily living in all such arrangements, as the only means by which they could be, even for a moment, upheld. Were the question to be settled upon a consideration of the respective moral merits of the two theories, I would say that the latter is greatly the preferable, as it implies a far grander view of the divine power and dignity than the other.'

So much for the supposed 'irreligious' tendencies of the 'Vestiges'! The question, after all, is, however, a purely scientific one, and must be settled by scientific men, upon scientific evidence, and wholly apart from its supposed bearings upon theological problems. From this merely scientific point of view, it may be said that the principal merit of the 'Vestiges' lay in the vigorous and successful attack which it made upon the doctrine of the 'special creation' of species.

When we turn, on the other hand, to the constructive side of the 'Vestiges,' we meet with few propositions peculiar to its author which would find an assured place among the generally accepted doctrines of modern zoology. It is, however, unnecessary to enter here into any further or more detailed analysis of the views put forward in this striking work. Our purpose will be sufficiently served by a brief notice as to the views of the author on the special question of the origin of 'species' among animals.

As regards the question of the fixity or mutability of 'species,' the author of the 'Vestiges' fully accepted Lamarck's views, in so far as he maintained that species are not invariable or constant. 'It is difficult,' he remarks, 'to regard the idea of species or specific distinction as descriptive of a fact in nature; it must be held as *merely applicable to certain appearances presented, perhaps transiently, to our notice.* The history of the question seems to be this. Naturalists, starting with a limited fund of observation—mainly, indeed, consisting of the remark which the most superficial observation supplies, that like usually produces like—lay it down as an axiom that species is a determined thing. In a little time, certain modifiabilities are observed. To maintain the axiom intact, these are called varieties. Afterwards, much greater variabilities are witnessed, even to the dissolution of genera among the cryptogams and cereals, and the community of algæ and fungi—water and land plants. Still, to keep the axiom whole, these are held in doubt, or relegated to a place in the elastic region of the varieties. Such is the stage which we have now attained. But this is a process the reverse of philosophical: it is to start with a theory, and then make the facts succumb to it. Were the process reversed, and the facts taken first, we should see that a great modifiability exists in organic nature, especially in the humbler departments of the two kingdoms. And seeing that this modifiability presents itself within the scope of a very limited experience, it might safely be inferred that something much greater would be detected, if our range of experience were extended, especially since the world presents us with results which can only be naturally accounted for in this manner.'

As regards, however, the *mode* in which species have originated, the writer of the 'Vestiges' rejects the views of Lamarck altogether, considering his theory as to the cause of varieties (and therefore of species) as 'so far from adequate to account for the facts, that it has had scarcely a single adherent.' In this, the writer of the 'Vestiges' does less than justice to Lamarck. The special theory of the French naturalist does not fail because it gives 'the adaptive theory too much to do.' It fails because it does not recognise how useful adaptations are preserved and strengthened. It was left to the genius of Charles Darwin to fill this all-important hiatus in the Lamarckian hypothesis.

The author entirely accepts the conception of a fundamental unity of organisation among animals; and regards this as implying 'that all were constructed upon one plan, though in a series of improvements and variations, giving rise to the special forms, and bearing reference to the conditions in which each animal lives.' He points out that this underlying unity of organisation is of itself a strong *à priori* argument against the idea of the separate creation of species. 'Organisms,' he remarks, 'we *know* to have been produced, not at once, but in the course of a vast series of ages; here we now see that they are not a group of individually entire things accidentally associated, but parts of great masses, nicely connected, and integral in their respective totalities. Time, and a succession of forms in gradation and affinity, become elements in the idea of organic creation. It must be seen that the whole phenomena thus pass into strong analogy with those attending the production of the individual organism.'

This last sentence leads us to the special theory proposed by the author to account for the existence of the present forms of life. This theory, termed by the writer the theory of 'progressive development,' may be stated in his own words.

'The several series of animated beings, from the simplest and oldest up to the highest and most recent, are, under the providence of God, the results, *first*, of an impulse which has been imparted to the forms of life, advancing them in definite times by generation through grades of organisation terminating in the highest dicotyledons and Vertebrata, these grades being few in number and generally marked by intervals of organic character which we find to be a practical difficulty in ascertaining affinities; *second*, of another impulse connected with the vital forces, tending in the course of generations to modify organic structures in accordance with external circumstances, as food, the nature of the habitat, and the meteoric agencies, these being the 'adaptations' of the natural theologian. We may contemplate these phenomena as ordained to take place in every situation and at every time where and when the requisite conditions are presented—in other orbs as well as in this—in any geographical area of this globe which may at any time arise—observing only the variations due to difference of materials and of conditions.'

Put briefly, the theory of progressive development is that the primitive cells, which constituted on this hypothesis the original forms of life, and which had been presumably produced by spontaneous generation, were advanced 'through a succession of higher grades and a variety of modifications,' in obedience to some law of an absolute nature, the whole process being analogous to the embryonic development of an individual animal. Just as each individual animal passes through a series of changes during the course of its development— these changes taking place in a fixed order—so the writer of the 'Vestiges' supposes that the primordial forms of life also passed through a series of developmental changes, the different stages of their development being represented by the life-assemblages of the successive great geological periods. These developmental changes are supposed to have taken place in a fixed order, and to have been progressive in character; and the present forms of life are supposed to represent the final term in the developmental cycle of these hypothetical primordial cells. It is not necessary to enter here into any discussion of the theory of progressive development. The obvious objection that an evolutionist of the Spencerian school would take to it is that, from his point of view, 'the impulse which has been imparted to the forms of life,' and to which their subsequent progressive development is supposed to be due, is a mere metaphysical conception, a hypothetical and scientifically inadmissible agency.

THE THEORY OF NATURAL SELECTION.

CHARLES DARWIN.

It remains to consider very briefly the leading points involved in the theory of 'the Origin of Species by means of Natural Selection,' which the world owes to the genius of Charles Darwin, and by which the entire science of zoology has been fundamentally altered. There is, indeed, no revolution so great as that effected by the introduction of a new *principle*; since that involves a reconstruction from the foundation upwards, and implies a much more serious change than the mere putting on of a roof, or the addition of a buttress or of any sort of pendicle, however important such may be in itself. Darwin, however, introduced a novel principle into biology; and in so doing he profoundly altered the entire attitude of naturalists and botanists towards the world of living beings. Moreover, when the organic world came to be viewed in the light of this new principle, it became at once evident that its complexities depended, to a large extent at any rate, upon causes which are open to our investigation, and are not wholly beyond our comprehension. The theory of the origin of species by special creation laboured under the inevitable defect that it 'closed the record,' and in many directions shut the door to further research. The theory of the origin of species by means of natural selection has not only brought to light a whole series of problems, many of which are of a most far-reaching character, but it has solved some of them, and has pointed out to us the way in which others may yet be solved at some future date.

As has been seen, the theory that the present state of the natural world was the result of its evolution from a former state did not originate with Darwin. Like others of the profoundest conceptions of the human mind, it had been more or less clearly recognised by more than one earlier philosopher, and notably by Erasmus Darwin and Lamarck. The theory that the 'species' of animals and plants now in existence had been produced by the modification of pre-existing forms of life, and that species were therefore not immutable, also did not originate with Darwin. Lamarck had definitely promulgated this theory, and other writers—such as Erasmus Darwin and Goethe—in the early part of this century or the close of the last, had put forth similar ideas. Lamarck's views, however, had remained little more than a barren speculation—unheeded by most, and scoffed at by many—and no change had been produced in the generally accepted views as to the nature of 'species' by the publication of the 'Philosophie Zoologique.' To Darwin is incontestably due the pre-eminent merit of having established a theory which satisfactorily explains the *method* in which species have been produced by evolution from other previously existing forms. No naturalist at the present day, it may safely be said, doubts that the theory of the origin of species by means of natural selection is true so far as it goes, and that it satisfactorily explains the principal difficulties which it can be legitimately called upon to explain. 'Natural Selection' is, in other words, universally recognised as a *vera causa*. The chief point that can be said now to be at issue among naturalists is not whether it be a genuinely active cause, but only as to the extent to which it can be *applied*—some regarding it as the sole factor in the production of 'species,' while others look upon it as being only one of many concurrent factors.

Darwin's life need only be referred to here in the briefest way, and only for the purpose of showing how thoroughly it qualified him for the task of elaborating and establishing his great theory. Charles Darwin was born at Shrewsbury, on the 12th of February 1809. His father was Dr Robert Waring Darwin, a physician of Shrewsbury, and his grandfather was the celebrated Dr Erasmus Darwin, whose life and writings have been previously noticed. At sixteen years of age, Charles Darwin went to Edinburgh to study medicine; but he soon made up his mind that the pursuit of medicine as a profession would not be in accordance with his tastes, and he accordingly betook himself in 1828 to Cambridge, with a view to studying theology. The influences of the place, however, combined, we may presume, with his own unconscious bent and aptitudes, soon had the effect of so far awakening his early love of nature, that he ultimately threw himself almost entirely into scientific studies. This result was also in large part due to the intercourse which he enjoyed with Professor Henslow, the well-known botanist.

In 1831, Darwin graduated as Bachelor of Arts, and in the autumn of the same year his final life-course was determined for him by his appointment to the unpaid post of naturalist to the *Beagle*, a ten-gun brig, commanded by Captain (afterwards Admiral) Fitzroy, and then under orders to proceed on a long surveying voyage round the world. This cruise occupied five years of Darwin's life, and constituted 'the real great university in which he studied nature, and read for his degree.' * During this memorable voyage, he not only collected a vast amount of scientific material of all kinds, but he accumulated an endless store of observations which might, and ultimately did, serve as the groundwork for his *magnum opus* on the Origin of Species.

In October 1836, Darwin landed at Falmouth, after his long and profitable cruise in the *Beagle*. The next three years were spent by him in London, his hands being fully occupied with preparing his journals for publication, and in making the needful editorial arrangements for the description of the great scientific collections which he had brought home with him.† By the advice

* Grant Allen, 'Life of Charles Darwin.'
† Darwin's 'Journal of Researches into the Geology and Natural History of the various Countries visited by H.M.S. *Beagle*' was published in 1839. The descriptions of the scientific collections were ultimately published in 'The Zoology of H.M.S. *Beagle*,' which appeared in 1840-44. In this magnificent work, the fossil mammals were described by Owen, the living mammals by Waterhouse, the birds by Gould, the fishes by Jenyns, and the reptiles by Bell.

of his friend Sir Charles Lyell—advice which his freedom from pecuniary necessities fortunately enabled him to take—Darwin, on his return home, sought no official scientific appointment. In 1839, he married his cousin, Miss Emma Wedgewood, and finally established himself at Down House, near Orpington, in Kent, which continued to be his home to the end of his life.

After his long voyage in the *Beagle*, Darwin never left England again, not even to pay a brief visit to the Continent. From his settlement at Down in 1839 onwards, he lived a quiet unostentatious life in his own home, unremittingly occupied with his scientific pursuits. On the 18th of April 1882, the great naturalist was attacked by sudden illness, and at four o'clock in the afternoon of the next day he breathed his last. He was buried in Westminster Abbey, in the presence of most of the foremost representatives of science in Britain; and his death deprived the scientific world of the most prominent figure that this generation has seen.

With regard to the vast mass of scientific work which Darwin produced, nothing further can be attempted here than merely to mention the titles of his larger works. His 'Journal' of researches made in the voyage of the *Beagle* was, as we have seen, published in 1839. Other fruits of the long series of observations which he made on the same voyage were published later under the names of 'The Structure and Distribution of Coral-Reefs' (1842), 'Geological Observations on Volcanic Islands' (1844), and 'Geological Observations on South America' (1846). Many of Darwin's geological observations (such as those on cleavage and foliation, on the structure of the 'pampas' of South America, and on volcanic islands) are of the highest importance and of permanent value; and his theory of the Origin of Coral-reefs obtained a world-wide reputation. Darwin, as previously mentioned, also edited the 'Zoology of the Voyage of the *Beagle*.' Subsequently to his return to England, he engaged in special zoological researches, and published his classical 'Monograph of the Cirripedia,' printed by the Ray Society in 1853; with a companion volume on the fossil species of the same group, which appeared under the auspices of the Palæontographical Society. In 1859 appeared the first edition of the 'Origin of Species by means of Natural Selection,' which rendered his name at once famous over the whole civilised world, and which gave rise to more discussion than perhaps has ever been produced by any other scientific book whatever. This work has been translated into almost all European languages, and the English edition now generally used is the sixth, published in 1872. Among the works which proceeded from the pen of Dr Darwin during his later years may be enumerated 'The Fertilisation of Orchids' (1862); the 'Variations of Animals and Plants under Domestication' (1867); 'The Descent of Man and Selection in Relation to Sex' (1871); and 'The Expression of the Emotions in Man and Animals' (1873).

The great principle which Darwin established in connection with the highly complex problem of the Origin of Species, is what is known as 'the Theory of Natural Selection, or the Preservation of Favoured Races in the

Struggle for Life.' Mr Alfred Russell Wallace has a conjoint claim to the discovery of this principle, as he published similar views to those of Mr Darwin in a memoir entitled 'On the Tendency of Varieties to depart indefinitely from the Original Type,' which appeared in the Journal of the Linnean Society in 1859, in the same year as the first edition of the 'Origin of Species' was given to the world. It is, as has been seen, an error to regard Mr Darwin as the originator of the theory of *Evolution*, as applied to animals and plants. It is the 'Theory of Natural Selection'—a theory which explains *how* evolution has taken place—with which his name will be always associated; and it is this theory alone of which we propose here to give a general outline.

The bases of the 'Theory of Natural Selection' may be laid down in the following propositions:

(1) The first proposition in the Theory of Natural Selection embraces what has been called the 'Malthusian law of increase'—the law, namely, that all living beings tend to increase more rapidly than their means of subsistence. The tendency of living beings, in fact, is to increase in a geometrical ratio, and this is true not only of all animals but also of all plants. In support of this law it is not necessary to take the cases of animals so prolific as the cod, the female of which produces annually about ten millions of ova; for the same law is exemplified quite as well by the elephant, which is considered to be the slowest breeder of all animals. Upon this point Darwin has made an interesting calculation. The elephant begins to bear young

at thirty years of age, and continues to produce offspring till it is ninety years old, during which time it has six young ones. The average age of the elephant may be calculated at about one hundred years, though this is often exceeded. On this basis, Darwin calculates that at the end of about seven hundred and fifty years the offspring of the first pair of elephants would amount to about nineteen millions of then living individuals.

(2) In consequence of this geometrical rate of increase among all living beings, it necessarily follows that there arises a 'Struggle for Existence' among animals and plants. Each organism fills a certain place in the world of nature, occupies a particular area, feeds on a particular kind of food, requires, in short, a particular set of conditions. As, however, every kind of animal and plant is constantly bringing into the world more young than can be accommodated, or for which suitable food can be provided, it follows that there arises among the young of each species a *competition*, a struggle both for a proper place and for proper food. This competition, which is seen quite as much in plants as in animals, is what is understood as the 'struggle for existence.' In using this term, Darwin premises that he does so 'in a large and metaphorical sense, including dependence of one being on another, and including (what is more important) not only the life of the individual, but success in leaving progeny. Two canine animals, in a time of dearth, may be truly said to struggle with each other which shall get food and live. But a plant on the edge of the desert is said to struggle for life against the drought, though more properly it should be said to be dependent

on the moisture. A plant which annually produces a thousand seeds, of which only one on an average comes to maturity, may be more truly said to struggle with the plants of the same and other kinds which already clothe the ground. The mistletoe is dependent on the apple and a few other trees, but can only in a far-fetched sense be said to struggle with these trees, for, if too many of these parasites grow on the same tree, it languishes and dies. But several seedling mistletoes, growing close together on the same branch, may more truly be said to struggle with each other. As the mistletoe is disseminated by birds, its existence depends on them; and it may metaphorically be said to struggle with other fruit-bearing plants, in tempting the birds to devour and thus disseminate its seeds.'

(3) The third proposition of the theory of natural selection is that all living beings are subject to variation. As has been previously seen, the individuals which compose any and every 'species' of animals and plants are not *precisely* alike. They invariably differ from one another in more or less numerous points, some of the differences being extremely minute, while others may be very conspicuous. We do not know whether variation is indefinite, and affects *every* part of the organism, or whether it is definite and is confined within certain limits. Nor has it been clearly proved whether variation is fortuitous, or whether it takes place in obedience to some determinate law, which governs the direction which it follows. It is, however, certain that 'variation,' to a greater or less extent, is of universal occurrence among all living beings.

(4) Some of the variations which occur in the individuals composing any species, are favourable to the species; some are unfavourable. That is to say, some variations will either help the individual to obtain more food, or to keep himself warm, or render him less liable to fall a prey to his natural enemies, or will otherwise help him in the struggle for existence. On the other hand, some variations will keep the individual back in the race for life, and will increase the difficulty which *all* individuals have in maintaining their existence. It follows from this, that in any given species of animals or plants those individuals which are born into the world in the possession of any favourable variations are, *cæteris paribus*, likely to be preserved; while those having unfavourable variations are likely to go to the wall and to be stamped out.

This law is what Mr Herbert Spencer has called the law of the 'Survival of the Fittest,' or what Mr Darwin has called 'Natural Selection.' This last name is in allusion to the fact that the action of 'Nature'—that is, the aggregate of natural forces—is to insure the 'selection,' out of the young of any species, of all those individuals which are 'fittest' for their surroundings. These young are preserved, while those not possessing any such favourable variations, and therefore not so well fitted for their surroundings, are weeded out and perish. The operation of the law may be illustrated by the imaginary example of the Giraffe, which Mr Darwin has himself used as illustrating the action of natural selection, and which was previously taken as illustrating Lamarck's view as to the action of external conditions upon the structure of animals. If we suppose, namely, that the giraffe, to begin with, possessed a neck of no more than normal length, and lived principally upon the ordinary terrestrial herbage; and if we further suppose a severe and protracted drought to occur in the region inhabited by the giraffes, we may assume that many individuals would perish for want of food, but that some would manage to survive. In all such cases there must be some general reason to account for the survival of the few who did survive, in preference to the many who perished. In this particular instance we may suppose that the individuals who survived were those who possessed necks of a slightly greater length than the average, and who, therefore, were better fitted for browsing upon shrubs or trees, after the herbage had been destroyed by the drought, than were the more normal individuals. This imaginary example, then, will show how the possession of a favourable variation tends to preserve certain individuals, in preference to those which are without the variation.

(5) But, the young of all animals and plants tend to *inherit* the peculiarities of their parents. Hence, favourable variations or peculiarities which preserve alive certain individuals of each species, will tend to be handed down to their offspring. On the other hand, individuals not possessing these favourable variations, or possessing unfavourable variations, are killed off, and do not have the opportunity of transmitting *their* peculiarities to offspring. The general action of the law of 'Survival of the Fittest,' or of 'Natural Selection,' is, thus, *to preserve all favourable variations* which may occur among the indi-

viduals composing any species, and *to destroy all unfavourable variations amongst the same.*

To use once more the imaginary illustration above employed, the longer necks which enabled certain individual giraffes to survive a drought, would be handed down by inheritance to their young. On the other hand, the comparatively short-necked individuals would not have the chance of leaving offspring because, by the hypothesis, they would be killed off.

Moreover, in the course of this transmission, the favourable variation (whatever it may be) will tend to become *intensified* in each succeeding generation, so long as the conditions which render the variation favourable to the life of the individual remain in existence. So long as this continues, the same process of 'selection' will go on in each succeeding generation; and the varying character will become in each generation successively stronger and stronger. Thus, in our illustration, so long as the region tenanted by the giraffe continued subject to periodic droughts, and so long as it was, therefore, good for the individual giraffe to have a long neck, the individuals in each generation which had the longest necks would have the best chance of survival. The best chance of survival, however, implies the best chance of leaving offspring, and in this way the neck of the giraffe might go on getting in each generation longer and longer, by the preservation of the individuals which possessed this variation to the greatest extent, and the elimination of those with shorter necks.

By means of this process of 'natural selection,' it is easy to comprehend how 'varieties' might be produced.

Nor can it be reasonably doubted that in the case of animals this *is* the process by which varieties are originated and established. But it has been previously seen that 'species' and 'varieties' pass into one another by imperceptible gradations. It is, in fact, impossible to lay down any fixed rule for the determination of where a 'variety' ends, and where a 'species' begins. If, therefore, it be admitted that 'varieties' are produced by 'natural selection,' it is not possible to deny that the same cause must have given rise to at any rate *some* of those groups of individuals which naturalists call 'species.' If this be conceded, it is an inevitable logical conclusion that *all* species have been thus produced by 'natural selection.' At any rate, the admission that *any* species have been produced by the operation of 'natural selection,' throws upon those who deny the universal operation of the law, the burden of proof that any particular species has *not* been produced by the action of the same law.

The above may be taken as a brief statement of the principal propositions upon which Darwin based his celebrated theory of the Origin of Species by means of Natural Selection. This statement would, however, be incomplete without a short additional exposition of what Mr Darwin has called 'artificial selection.' In the case, then, of our domestic animals and their innumerable varieties, there is the obvious fact that the law of 'natural' selection is prevented from operating in its entirety owing to the action of man. Man, in the case of his domestic animals, steps in as a *deus ex machinâ*, and more or less efficiently interferes with the law of natural selection by protecting certain individuals of a species in the struggle for existence, and affording them assistance which they could not have had in a wild state. The individuals of the Wild Boar, for example, have to face the rigid and merciless operation of the law of natural selection, and the weakest therefore go to the wall. The individuals of the Domestic Pig—the same animal really as the wild boar—are so far relieved from the action of the law of natural selection, that man feeds them when they are hungry, protects them from the cold artificially, and, so far as he can, cures them when they are ill. Man does of course the same thing to the weaker individuals of his own species, and all such things as poor-laws and the like are, in the Darwinian sense, attempts on the part of man to defeat or neutralise the operation of 'natural selection.' In the case of many varieties of our domestic animals, it is certain that man's interference has gone so far as to render them wholly incapable of facing the law of natural selection in its untempered severity. In other words, there are many of our domesticated breeds of animals which would infallibly become exterminated if they were turned loose to make their own living, and if the protecting hand of man were wholly withdrawn from them.

Man, however, not only protects domestic animals in this way from the direct action of surrounding conditions, but at the same time exercises, on his own behalf, a sort of 'selection,' analogous to 'natural selection,' but necessarily operating within much narrower limits, and also exercised in a much more arbitrary fashion. Darwin has given a masterly exposition of the whole of this subject under the head of what he has called 'artificial selection;' and a few words may be said here as to what he understands by this name. Our domesticated animals, as is well known, have in all cases originated from wild species, which have gradually been brought under the influence and dominion of man. The same is true of all our domesticated, or rather cultivated, plants. In certain cases—as that of the pig above referred to—we not only have the domesticated breed or breeds, but we are also acquainted with the wild species from which the domestic form was derived. In other cases, the domesticated animals have undergone changes so great that we can no longer point with certainty to the wild forms in which they originated. In some cases, it may be, the wild form is no longer in existence. In all cases, however, our domestic animals show, more or less conspicuously, two remarkable characteristics or tendencies. One of these is that they exhibit more numerous and more marked 'varieties' than is the case, as a rule, with wild species. They have a more pronounced *tendency* to variation than wild animals have, and their variations also extend through a wider range. The other is, that the peculiarities which are distinctive of our domestic animals as compared with their wild forms, are not of such a nature as to fit the animal better for its natural wild life, but, as specially insisted on by Mr Darwin, are adaptations to the taste, or fancy, or requirements of man. Thus, any modifications produced by natural selection in the wild boar would be in the direction of making it stronger, or enabling it better to resist cold, or rendering it fitter to cope with its natural foes, or the like. Man, however, does not desire any improvements of this kind in the domestic pig. He does not, for example, wish to increase its muscular power and consequent activity; because he wants it to fatten readily, and vigorous exercise tends to keep an animal lean. What is true of the pig is true of all our domestic animals, though it is more evident in some than in others. The variations, for instance, which separate the different breeds of the dog from one another are exceedingly well marked, and they are all variations which adapt particular breeds for the special purposes for which man wants them. On the other hand, the different breeds of the Goose differ little from one another, or from their wild form (the Grey Lag Goose), because man's demands from the goose are few and simple, and are quite well answered by the ordinary form of the species.

The causes of the above-mentioned peculiarities of domesticated animals, as compared with wild ones, have been fully expounded by Mr Darwin, and are readily intelligible. As regards the first of them—namely, the tendency to excessive variation shown by domestic animals —the cause is to be found in the varied character and artificial nature of the conditions under which they live. Wild animals are exposed, as regards each species, to an approximately uniform and unvarying set of conditions, and the conditions are alike for all the individuals of the species. Variation does not become excessive, because the tendency of natural selection is to destroy all variations which are not good for the *individual* itself in its natural condition. On the other hand, domestic animals are kept by their masters under very different sets of conditions, as regards different individuals of the species, and man at the same time prevents the law of natural selection from rigidly exterminating those individuals which happen to be born with variations which would be hurtful to the species in a wild state. The fact that domesticated animals exhibit peculiarities which are in no way adaptations to their natural surroundings, but which are mere adaptations to man's wants or tastes, is explained by 'artificial selection.' Man, namely, has as regards each domestic animal an ideal of what he wants. It may be that he has no consciousness of having any such ideal before him, but it may be taken as certain that he possesses it nevertheless. 'Artificial selection' consists essentially in the choice which man exercises as to the young of his domestic animals, in respect to which he will allow to live, and which he will destroy. In the case of the young of each of his domesticated animals, a man sees some individuals having peculiarities which he thinks will be useful to him, or which come nearest to the ideal which he has formed of the animal, or of what the animal *ought* to be. Such individuals he keeps, and permits to have offspring; so that the peculiarities which induced him to keep these individuals are perpetuated and handed down to future generations, becoming in the process intensified. On the contrary, all those individuals amongst the young, which do not conform to man's ideal standard of perfection, are either killed off on the spot, or are, at any rate, prevented from leaving offspring behind them. In this way, by a long-continued process of *selecting* the particular individuals which he will allow to live and to breed, man has succeeded in producing the numerous domesticated varieties of animals. In the case of savage tribes of men, this selection is no doubt carried on unconsciously, but among the breeders of cattle, or among pigeon-fanciers, it is a strictly scientific process, carried on consciously and deliberately, and according to rules, which are none the less fixed that they are largely 'rules of thumb.'

Those who wish to understand this most interesting subject in all its bearings must turn to the pages of the 'Origin of Species,' where it is fully treated by the hand of the master. All that need be done here is to say one word as to the relation between the known facts of 'artificial selection' on the one hand and the theory of the origin of wild species by 'natural selection' on the other hand. If it be admitted, namely, that our numerous varieties of domesticated animals owe their peculiarities to the 'selection' exercised by man during the comparatively brief period during which he has existed upon the earth, it is not unreasonable to suppose that 'natural selection,' operating through an infinitely longer period, and by methods much more subtle and far-reaching, has produced the different wild 'species' of animals by modifications of one or more aboriginal types. The unquestionable facts, therefore, as to the production of our domesticated breeds of animals from wild species by means of 'artificial selection,' afford a strong presumption in favour of the theory that our existing wild species have been produced by the modification of pre-existing wild species through the operation of 'natural selection.'

THE THEORY OF NATURAL SELECTION
(CONTINUED).

HAVING now given the briefest possible sketch of the Theory of Natural Selection, as expounded by Mr Darwin, it may be well to notice, with equal brevity, the leading objections which have been urged against this theory by various naturalists, and notably by Mr Mivart.* It may also be as well to enumerate shortly the chief *general* grounds upon which naturalists base the now generally accepted belief that species have been produced from preexisting species by the action of *some* law of evolution, apart from the question of the method or methods in which this law operates.

Numerous difficulties admittedly have to be met, if we attempt to apply the theory of natural selection (even when combined with what Darwin has called 'sexual selection') as the sole principle involved in the production of 'species.' Many of these difficulties are of a special nature, affecting special cases only, and they need no discussion here. It is possible that many of these special

* *The Genesis of Species*, by St George Mivart, 1871.

difficulties may disappear in the light of wider knowledge. There are, however, certain general difficulties which demand a moment's consideration, as indicating that though we admit the action of 'natural selection' to the full, we must nevertheless look beyond and outside this for the *complete* explanation of the existence and origin of species. The general difficulties in question were perfectly recognised by Mr Darwin, and have been met by him, as far as it is at present possible to meet them. The principal are the following : *

(1) One of the most general, and certainly one of the most serious of the difficulties in the way of the theory of natural selection is 'the uselessness of many organs in their incipient stage.' Hosts of structures (such as the milk-glands of the Quadrupeds, or the whalebone plates in the mouth of the Whalebone Whales) are exceedingly useful to the animal when perfectly developed; but it is inconceivable that they could have benefited the animal when first they began to be developed. According to the theory of the evolution of species in general, and the theory of natural selection in particular, milk-glands did not exist in the animal forms out of which the class of the Mammals was evolved, nor did baleen-plates exist in the ancestors of the Whalebone Whales. There must, therefore, have been a time when milk-glands and baleen-plates respectively first came into existence, and it is impossible to suppose that they were suddenly produced in complete structural and functional perfection as we now see them. On the contrary, they must, to begin with, have been mere

---

* An excellent résumé of these objections is given by Mr Pascoe in his *Notes on Natural Selection and the Origin of Species*, 1884.

rudimentary structures, functionally useless, and it can only have been in the course of development during many successive generations, that they assumed their present perfection. Now there is absolutely no evidence to show that the fine beginnings of structures can be useful or profitable to the animal possessing them. They may be harmless, but that is all that can be said. It is, however, the very essence of the theory of natural selection, that the law of the struggle for existence is powerless to preserve or intensify any structures except such as are *useful to the individual*. The fact that a structure may be useful to the *race* is not enough, as final causes or ends are wholly excluded from the theory of natural selection. Upon the whole, the difficulty of accounting for the preservation of incipient organs and structures by the action of natural selection appears to constitute the most formidable of the arguments which have been urged against Mr Darwin's views; since it is a general difficulty, and strikes at the very root of the theory of natural selection.

(2) A second general objection of great weight is that unless 'many individuals should be similarly and simultaneously modified,' there would be little chance of any useful variation which might have appeared in a species being ultimately preserved and handed down. Any new structure or organ, or any alteration in a pre-existing structure, must be slowly produced, and pass through an incipient stage. If, however, such a new structure, or alteration in an old structure, appeared, to begin with, in only one or two individuals of a species, it could hardly be preserved, as it would be 'lost by subsequent intercrossing with ordinary individuals.' But it is hardly

probable that any variation would simultaneously appear in many individuals of a species; and we have at any rate no evidence to show that this ever occurs.

(3) The theory of the origin of species by means of natural selection, in the third place, implies that the production of any given species from any pre-existing species can only be effected by gradual modification, and therefore through the intervention of a long series of intermediate or transitional forms. Moreover, the transitional forms by which we should pass from a given species to the pre-existing species from which it was developed, must, on the theory of natural selection, be so closely related to one another as to render it difficult to distinguish them. In other words, if we had before us all the forms by which one species had been gradually converted into another, we should not have the slightest difficulty in recognising the distinctness of the individuals forming the extreme terms of the series; but the individuals standing between the extremes would pass into one another by such fine gradations as to render their separation almost or quite impossible. It seems also clear that, in the modification of any one species into any other, the total number of the individuals of intermediate or transitional form must greatly exceed the total number of individuals contained in the original species and the new species put together. Now, if all species of animals, living and extinct, have been produced by gradual modification from pre-existing species, we ought to find abundant evidence of the existence of the infinite number of transitional forms postulated by the theory of natural selection. In fact, as these transitional

forms must have greatly exceeded in total number the combined number of individuals which are clearly recognisable as distinct species, we ought to find *more* abundant evidence of their existence than of the existence of the separate species. As a matter of fact, however, the study of extinct animals does not afford more than very incomplete evidence as to the existence of the numerous and closely graduated transitional forms required by the theory of natural selection. It is true that palæontology has brought to light many forms of animals which are distinctly intermediate in their characters between groups which would otherwise stand far apart. Thus, we have numerous extinct types which bridge over the gap between the reptiles and the birds; and others which stand intermediate between the existing horses and their original five-toed ancestors. So far, then, palæontology unquestionably lends support to the general theory of the evolution of species from pre-existing species. The theory of natural selection requires, however, more than this. It requires that there should be a *series* of intermediate types graduating into one another by slight and hardly perceptible differences. In some cases, as regards allied species of animals, such a continuously graduated series can be shown to exist (in some extinct Shell-fish, for example). In most cases, however, it must be admitted that palæontology has so far failed to demonstrate the past existence of the numerous and finely-graduated series of transitional forms between different species absolutely demanded by the theory of natural selection. Such transitional forms as are known for the most part stand quite sharply distinguished from one another and from the types which they connect. Mr Darwin has met this difficulty by pointing to the great 'imperfection of the palæontological record,' the fossil forms known to us doubtless forming only an insignificant fraction of those which once existed. This argument is entitled to receive great weight; but it does not sufficiently account for the *general* absence of graduated intermediate forms. This, however, is a point which cannot be further discussed here, and upon which each investigator will decide, in one sense or the other, according to the particular direction in which he may be led by his studies.

(4) It is, again, assumed upon the theory of natural selection, that 'variation' among the individuals of a species is *indefinite*, both in amount and direction. It would appear that the theory of the origin of species by means of natural selection requires a belief in the 'omnifarious' nature of individual variation. The action of 'Natural Selection' would, of course, still go on, even supposing variation to be strictly limited in amount; but in this case it is hardly conceivable that our existing species should owe their origin to natural selection, as the principal or sole factor in their production. On the contrary, it seems necessary to suppose that variation affects, or may affect, all parts of the organism, and that there are no limits to the extent of its operation, though the single steps of the process are small in amount. We have, however, no positive evidence which would enable us to assert, as a scientific fact, that variation is thus omnifarious and indefinite. The evidence actually in our possession is admittedly small, because it only

extends back to the beginning of the human period; but, so far as it goes, it would rather support the view that variation is limited and definite both in amount and direction. The 'artificial selection,' for example, which man has exercised in the case of his domestic animals for some thousands of years, has not, so far, resulted in the production of a single new 'species.' New 'varieties' have been produced, but that is all; and we know that these *may* appear suddenly (as in the instance of the Black-shouldered Peacock), without the direct or indirect action of man at all. Besides, if variation be indefinite, it is difficult to account for the constantly-recurring phenomenon of the extinction of species—a phenomenon which is, on any hypothesis, very difficult to satisfactorily explain. So far as wild animals are concerned there is no direct evidence to show that a single 'species' has come into existence since the beginning of the historical period; nor is there any evidence to show that during the same period a single wild species has become extinct, except only where its extinction has been the result of the interposition of man.

The points above enumerated are sufficient to show that there are great difficulties in the way of accepting 'Natural Selection' as the *sole* agent in the production of species. That it is *one* agent, and an important one, is a matter that does not admit of doubt. Under any circumstances, however highly we may rate 'natural selection' as an agent in the production of species, it remains certain that we are still almost entirely ignorant of the causes of the two fundamental laws which have to do with the production of species—namely, the law of variation and the law of inheritance. Our ignorance as to both of these is freely and fully admitted by Mr Darwin. The theory of natural selection does not profess to explain *why* variations occur; it only explains how those variations which are useful to the individual are preserved, and how those which are injurious are 'rigidly destroyed.' Like all other hypotheses as to the origin of species, it leaves us entirely in the dark as to the *causes* of variability. The law of variation is therefore an unknown law, lying behind the law of evolution, and possibly beyond the limits of scientific investigation. Similarly, the laws of inheritance are almost wholly unknown. 'No one can say why the same peculiarity in different individuals of the same species, or in different species, is sometimes inherited and sometimes not so; why the child often reverts in certain characters to its grandfather or grandmother or more remote ancestors; why a peculiarity is often transmitted from one sex to both sexes, or to one sex alone, more commonly but not exclusively to the like sex' ('Origin of Species,' page 10).

That 'species' have originated by modifications through descent may now be taken as an accepted doctrine in modern zoology. It is Mr Darwin's supreme merit to have brought about this radical change in the views of naturalists by the establishment of the law of 'natural selection,' which for the first time rendered possible an explanation of the method in which the modifications of specific forms are caused. Whether or not natural selection has been 'the exclusive means of modification' is a point upon which different naturalists hold different opinions. Mr Darwin himself believed that it was at any rate 'the most important' means. Whatever may be the view ultimately adopted as regards this point, there is overwhelming evidence in favour of the belief in *some* general law of evolution, by which all animal and vegetable species have been produced. The evidence in favour of this may be briefly stated as follows:

(1) All the animals belonging to each great primary division of the Animal Kingdom are constructed upon one fundamental plan, which is capable of endless modifications, but is never lost. Thus, to give one example, the fishes, amphibians, reptiles, birds, and quadrupeds, which together constitute the 'sub-kingdom' of the Vertebrate Animals, are all built according to one common plan. However unlike they may be to one another in the details of their organisation, 'homologous' structures can be traced throughout the ground-plan of them all. This unity of plan in the types of life which compose each sub-kingdom is, however, inexplicable upon any other view than that it is the result of blood-relationship, and depends upon descent from a common ancestor, which possessed the essential structural characters distinctive of Vertebrates as a whole.

(2) The animals composing each sub-kingdom are constructed upon the same plan, and the 'sub-kingdoms,' taken as whole, stand therefore separate and apart. But there exist transitional forms by which one sub-kingdom is linked with another. Thus the singular marine animals known as the Sea-squirts (*Tunicata*) form a link between the true Shell-fish (*Mollusca*) and the Vertebrate Animals. In certain points, namely, in their

organisation, they approach the ordinary Shell-fish, while in others they show a relationship with the lower Vertebrates.

(3) It is a well-known embryological law that the young animal in the early stages of its development commonly possesses structures which it does not possess in its adult state. It is also a well-known law that structures which have only a temporary existence in the embryo of one animal, are often found existing throughout life in the adult of some other animal; and that when this occurs, the latter animal will occupy a lower position in the animal scale than the former. Thus, the embryo of the Quadrupeds possesses on each side of the neck a series of transverse slits or fissures (the so-called 'visceral clefts'), which lead down from the surface into the upper part of the gullet (the 'pharynx'). In the adult Quadruped no traces of these clefts are seen, only one of them remaining at all (the opening of the ear), and that only in a much modified form. On the other hand, the embryo of the Fishes not only possesses these clefts, but they are permanently retained, and are present therefore in the adult, in which they become connected with the gills. It seems, however, impossible to satisfactorily explain the possession of visceral clefts by the mammalian embryo, except upon the supposition that the Mammals and the Fishes alike have descended from a common ancestor in which these structures were present. The general fact, therefore, that the embryos of animals so often possess structures which are found in the adults of other animals, is strongly in favour of the belief in the production of animals by evolution from common ancestral types.

THE THEORY OF NATURAL SELECTION. 303

(4) This view is further borne out by the common existence in adult animals of what are known as 'rudimentary organs,' or, in other words, imperfectly developed organs which are of no use to their possessor. Thus, ordinary Snakes do not possess either the fore or hind limbs; but the Boas and Pythons possess rudimentary hind-limbs in the form of a pair of horny spurs. Again, the Whalebone Whales have no teeth; but they exist nevertheless in the young animal, though they remain buried in the jaw and never cut the gum. The same is true of the upper front teeth in Ruminant animals, which also do not cut the gum, and are therefore of no use to the animal. Another instance may be taken from the whales, which show no signs of hind-limbs externally, but which for all that often have the rudiments of these limbs hidden internally. The only satisfactory explanation of the general nature of rudimentary structures which it seems possible to give, is that they are structures which existed in a fully-developed condition in the remote ancestors of an animal, but which have gradually dwindled down in size and have lost their function through long-continued disuse. Sometimes rudimentary organs may be 'nascent' structures—that is, structures which in course of time may become functionally useful to the animal; or sometimes they may merely represent the atrophied condition of structures which the embryo possessed; but this does not affect the above general explanation. Accepting this view, we should judge that the whalebone whales were descended from some type of Mammal which possessed teeth in its jaws, and which was at the same time provided

with the hind-limbs as well as the fore-limbs. Similarly it would be concluded that the ancestral type of the Ruminants possessed well-developed upper front teeth; and that the snakes, though now footless, were descended from some reptilian type in which the limbs were present. Rudimentary organs, therefore, strongly support the view that the different forms of animals have been produced by modification from older and different forms.

(5) Lastly, the known facts of Palæontology offer the strongest support to the general theory of the evolution of animal forms from pre-existing species. Amongst extinct species we are constantly meeting with types which stand intermediate between groups otherwise more or less remote. One of the most famous examples of this is afforded by the fossil forms which link together the two groups of the Reptiles and the Birds—two classes of animals now so little resembling each other, that no one save a naturalist would ever suspect a relationship between them. Thus the past has yielded up to us the remains of true reptiles (the Deinosaurs) which walked upon their hind-legs, like birds; other reptiles (the Pterodactyles) possessed the hollow bones and the power of genuine flight characteristic of the living birds; some genuine birds (the *Odontornithes*), finally, resembled the Crocodiles in having the jaws furnished with numerous pointed conical teeth. Another famous example of the intermediate forms which palæontology has brought to light is that afforded by the extinct horse-like Quadrupeds of the Tertiary period. It is well known that our present Horse is peculiar in having only a single fully-developed toe on each foot. This toe corresponds with the middle

toe (or third toe) of an ordinary quadruped. If the skeleton of the horse's foot be examined, it will be seen that lying by the side of the great middle toe are two little splint-like bones, one on each side, which are the

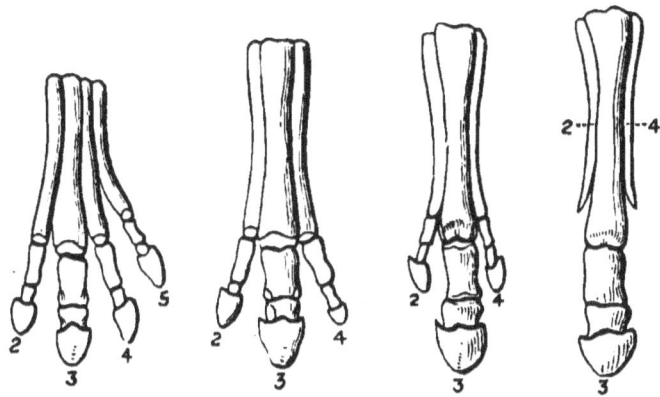

FEET OF FOSSIL EQUIDÆ.

'rudiments' of the index (or second) toe, and the ring toe (or fourth toe). The horse, therefore, possesses a foot with one complete toe and two incomplete ones; the outermost toe (the little or fifth toe), and the innermost toe (the thumb or great toe, or first toe) having no representatives at all. If, however, there be any truth in the general doctrine of evolution, it may be taken as certain that the horse has descended from a five-toed ancestor, since the typical Mammals possess five digits to the foot. Through the researches of Gaudry, Marsh, and others, it may now be confidently asserted that the horse *has* descended from a five-toed form. Thus, we meet with a number of horse-like animals, all now extinct, in which we find the foot, as we trace them backwards into the past, to

T

become progressively nearer and nearer to the normal pentadactylous type. In comparatively modern strata, we find the *Hipparion*, in which the two little splint-bones of the living horses are so far developed that they project externally and carry little hoofs at their ends. The foot is, therefore, three-toed in the *Hipparion*, but the animal still walked upon the great middle toe, and the lateral toes were functionally useless, as they did not touch the ground. In the still older *Anchitherium*, the two lateral toes are sufficiently developed to touch the ground, but the middle toe is still much the biggest, and is the toe upon which the weight of the body is principally supported. In the still older *Orohippus*, the fore-feet are four-toed, the fifth or little finger being now developed, but the thumb is still wanting, and the hind-feet have only three toes. Finally, in the *Eohippus*, the oldest type of equine animal yet discovered, the fore-foot possesses four complete toes, with a rudimentary thumb (or first toe) in addition, thus becoming morphologically five-toed. The above gives, of course, an exceptionally striking instance of how palæontology enables us to trace the *line of descent* of some particular living animal; but there are innumerable instances in which fossil forms exhibit characters which more or less extensively bridge over the gaps separating groups apparently widely remote. Upon the whole, therefore, the evidence of palæontology, though lending but a partial support to the theory of the origin of species by means of natural selection alone, is overwhelmingly in favour of a general theory of the evolution of animals from other pre-existent types.

INDEX.

Amphibia, classification, by Linnæus, 59.
Anaima, or animals without blood, 17.
Analogy, likenesses of, 173.
Anatomy, 8; Cuvier on, 140; value of Cuvier's contributions to the science of, 151.
Anchitherium, 306.
Anemones, sea, 190.
Animal kingdom according to Ray, 47; defined by Linnæus, 58; by Swainson, 175.
Ant-eater, spiny, 162, 166, 176.
Aptera, 60.
Aquatic animals, Ray's division of, 47.
Arachnida, the, of Cuvier, 146.
Arbutus, the, 214.
Arctic area, 215; plants and animals of, in Britain, 219.
Aristotle, the father of zoology, 6; *History of Animals*, 12; attitude towards natural history, 13; observations on cuttle-fishes, 14; first created a scientific method of treating the animal kingdom, 15; his division of animals, with and without blood, 17; scientific prosecution of natural history closed for centuries with his death, 18.
Armorican type of plants, 214; migration of, 221.
Articulata of Cuvier, 146, 147.
Artificial selection, as expounded by Darwin, 289, 291; production of domesticated breeds from wild species by means of, 292; has not resulted in the production of a new species, 299.
Ascidian molluscs, 139.
Ashmole, Elias, founder of Ashmolean Museum, 68.
Asturian area, 214; flora of, 221, 222.
Aves, or birds, classification of, by Linnæus, 59.

Bell, Professor Thomas, 186.
Belon, 21.

Bewick, Thomas, account of his life, 109.
Birds, 47, 59, 231; air receptacles of, 79; *Genera of*, by Pennant, 103; armed with spurs, 230.
Boar, the, 3, 101, 102, 230.
Box, the, 215.
British Museum, the, 64, 67; Sir Hans Sloane's bequest, 69.
Brongniart, Alexander, 153.
Bryony, the, 215.
Buffon on the possible evolution of species, 226; his *Histoire Naturelle*, 226; assists Lamarck, 240.

Carrion-crow, the, 50.
Catacombs, mummies found in, 157.
Cattiwake, the, 26.
Cetacea, 47, 49, 145.
Challenger Expedition, the, 212.
Chambers, Robert, and the authorship of the *Vestiges of Creation*, 264; the *Vestiges* more than an attempt to explain the origin of species, 265; conclusions set forth therein, 266-268; its chief merit the attack upon the doctrine of the special creation of species, 270; views of the author on the origin of species, 271; rejects Lamarck's views, 272; accepts conception of a fundamental unity of organisation among animals, 272; theory of progressive development, 273.
Charlton, William, the collection of, 68.
Chough, the, 50.
Classification in natural history, 9.
Cod, prolific nature of the female, 281.
Cœlenterate animals, 176, 190.
Conchology, researches in, 188.
Cooper, Sir Astley, on John Hunter, 83.
Coral (deep-sea) zone, 211.
Corallines, essay on, 95.
Crocodiles, 304.
Crow family, Ray's names for the, 50.

Crustacea, 47; Cuvier on, 146.
Cuckoo, 80.
Cuvier, 136-167; scientific papers, 138, 139; his great scientific activity, 142; work as a specialist in zoology, 142; on the Mollusca and fishes, 142; studies in palæontology, 143; treatise on the animal kingdom, 144; advances which Cuvier effected in zoological science, 145; studies of invertebrate animals, 146; vertebrate animals, 147; his grouping of the animal kingdom, 148, 149; recognised the true principle of philosophical classification, 150; contributions to morphology, 151; chief founder of science of palæontology, 152; investigations of tertiary rocks, 154; controversy as to origin of fossils, 154; a 'catastrophist,' 155; views on the origin of species, 155; on Lamarck's views, 157; gave a scientific basis to palæontology, 158; established the law of correlation of organs, 159; on the marsupials, 164.

Dalyell, Sir John Graham, 191.
Darwin, Charles, his birth and education, 277; cruise in the *Beagle*, 278; journal of researches, *note*, 278; scientific labours and death, 279; list of works by, 279, 280; established the theory of natural selection, 280; bases of theory, 281; variation, 283, 284; his illustration of the giraffe, 285; his theory of artificial selection, 287-292; objections to his theory of natural selection, 293.
Darwin, Erasmus, 223-235; intimately connected with theory of origin of species, 223; life of, 224; recognised natural variations and the principle of heredity, 227; on community of descent, 228; on probable cause of modification of species, 229; on transformations of animals, 229, 230, 231; concluded that all animals have a common origin, 233; taught the origin of species by descent with modifications, 234.
Deinosaurs, 304.
Development, science of, 8; doctrine of progressive, 264-274.

Distribution of plants and animals according to Forbes, 213, 214, 215, 216.
Dodo, stuffed specimen of, 68.
Donovan, Edward, and works by, 108.
Drury, the entomologist, 108.
Duck-mole, 162, 167, 176.
Duport, James, 22.

Eel, the electric, 79.
Elephant, supernatural qualities ascribed to, 3; trunk of the, 230; Darwin's calculations as to age, &c., 282.
Ellis, John, 93.
Embryology, 8, 151, 302.
Enaima, or animals with blood, 17.
Entomology, 189.
Eohippus, 306.
Equidæ, feet of fossil, 305.
Evolution, 10; the key to biological problems, 223; Lamarck's theory of, 223, 243; Buffon on, 226; facts of palæontology support theory, 304.

Fish-hawk described by Wilson, 128.
Fishes, 47; Willughby on, 43; Cuvier and Günther, 44; classified by Linnæus, 60; Cuvier's researches in the department of, 142.
Fleming, Rev. John, 184, 185.
Forbes, Edward, 184, 192-222; fondness for natural history, 193; college training, 195; abandons the study of medicine, 197; foreign travel, 197; dredging expeditions, 198; appointed naturalist to the *Beacon*, 199; professor of Botany in King's College, 200; appointed to the Natural History chair in Edinburgh, 202; premature death, 203; his contributions to natural history, 203, 204; studies of the British mollusca, 205; researches into the distribution of animals, 205; his belief in the fixity of species, 206; theory as to *genera*, 208; on the distribution of animals in the sea, 209; four zones of depth, 210; theory of the absence of life in deeper parts of ocean disproved, 212; studies in distribution of plants and animals, 213, 214, 215; theory of the distribution of plants and animals, 217; the three floras, 220, 221.

Forskal, 55.
Fossils, Ray's theory of, 31; Lister on, 91; controversy as to origin of, 154.
Fregilus genus, 51.
Frog-fish, the, 231.

Gastric juice, John Hunter on, 78.
Genera of recent and fossil shells, 188.
Gentian, the blue, 215.
Geographical distribution of animals, 8.
Geology and its relations to palæontology, 153.
Germanic type of vegetation, 216.
Gesner, 21.
Giraffe, the, Lamarck's theory as to, 252; an example of modification, 285.
Glacial period, migration of plants and animals during the, 218, 219.
Goethe on mutability of species, 227.
Goose, different breeds of, 290.
Graptolites, 58.
Gronovius, 53.
Gulf-weed, *note*, 222.
Günther, A., on Willughby's *Ichthyology*, 44; on Cuvier, 151.

Harris, Moses, 108.
Hasselquist, 55.
Hedgehog, Hunter's researches on, 81.
Heredity, 227.
Herring, migration of, 98, 99.
Hipparion, the, 306.
Hog, the, Pennant on, 100, 101; used as a beast of draught, 102.
Home, Everard, 77-84; first conservator of the Hunterian Museum, 88; destroys Hunter's unpublished observations, 89; publishes *Lectures on Comparative Anatomy*, 89.
Homology, likenesses of, 173.
Hooded-crow, 50.
Horse, descent from five-toed form, 305.
Hunter, John, founder of science of comparative anatomy, and of Hunterian Museum, 69-76; list of his philosophical and zoological papers, 78, 79; correspondence with Jenner, 80; researches upon the cuckoo, 80; on hedgehog, 81; work on the blood, inflammation, and gun-shot wounds, 82; death, 85; personal appearance, 86; distinguishing character of his mind, 86; museum his *magnum opus*, 86, 87.
Hunter, Dr William, a great anatomist, 71; his extensive museum, 72; excellence as a lecturer and demonstrator, 83; controversy with John Hunter, 84.
Hunterian museum, 64, 69, 72, 86, 87; bought by government and attached to the College of Surgeons, 88; descriptive catalogues of, 89.
Hydra, the, 94.
Hydra-tuba, 191.

Ichthyornis, 162.
Insects, 47, 60, 61; Ray on, 35; Willughby on, 45; classification of, by Linnæus, 60; Cuvier on, 146.
Invertebrates, Cuvier on, 146, 147; Lamarck on, 241.

Jackdaw, the, 50.
Jardine, Sir William, 128, 184.
Jenner, Dr, a pupil of John Hunter, 77; Hunter's correspondence with, 80.
Johnston, Dr George, 190.
Jussieu, Antoine de, 54; Bernard de, 94; botanical classification by, 238.

Kangaroo, pelvis of, 163; jaw of, 164.
Kentish Flora, the 221.
Kirby, Rev. William, 189.
Kittiwake, the, 26.
Koster, Mr, travels in Brazil, 170.

Lamarck, Chevalier de, 236-263; on the origin of species, 155; order of his botanical classification, 239; theory of evolution, 243; saw the difficulty of separating species from varieties, 246; rejected the idea of constancy of species, 247; views on variation, 249; believed simpler forms of life were evolved first, 250; agencies concerned in modification, 251; thought that external conditions had evoked a corresponding structure, 252; instance of the giraffe, 252; climatic differences in animals, 256; recognised that the habitual use of an organ leads to its corresponding growth, 262.
Laminarian zone, 210.

Latham, Dr John, 107.
Lewes, George Henry, 12-17.
Linnæus or Linné, Karl von, 46-52; publication of *Systema Naturæ*, 53; visits England, 53; scientific labours, 55; position as a botanist and zoologist, 56; his classification the simplest and most complete up till that time, 57; division of natural objects into kingdoms, 58; great merits of his classification, 61; introduction of binomial system, 62; indications of a farther expansion of the Linnean nomenclature, 63.
Lister, Martin, 33, 90.
Littoral zone, the, 210.

Macgillivray, William, 187.
Macleay, W. S., 168.
Madder, wild, 214.
Maillet, Benedict de, 156.
Malacia, 47.
Malacostraca, 47.
Malthusian law of increase, 281.
Mammalia, division of, by Linnæus, 58; primates, 58; least satisfactory features of this division, 59; Cuvier on osteology of, 151.
Mammals, 44, 47, 49.
Manatees, 49.
Marsupial quadrupeds, structure of, 164; where found, 164; fossil, 166.
Mastodon, 143.
Megalonyx, 143.
Migration, animals and plants, 217-219.
Mivart, St George, objections to law of natural selection, 293.
Mollusca, 47, 91, 187; Cuvier's researches on, 139, 142, 147, 151; Forbes on, 204, 205.
Montagu, George, 107.
Morphological likenesses, 173.
Morphology, 8; 151.
Mullein, the, 215.
Museums, great, 64-89.

Natural History, beginnings of, 1-6; no system previous to Aristotle, 7; only the aggregate history of all known species of animals, 7; departments known as morphology or anatomy, physiology, embryology, and geographical distribution of animals, 8; of palæontology, 9; of taxonomy, 9; of evolution, 10; of psychology and teleology, 10; scientific prosecution of, came for centuries to a close at the death of Aristotle, 18; contributions of Pliny the Elder to, 18; awakening in seventeenth century, 18, 21; artificial and natural classification, 47.
Natural Selection, theory of, 275; has brought to light a whole series of problems, 276; first established by Darwin, 277, 281; bases of the theory of, 281; operation of this law in the case of the giraffe, 285; preserves all favourable variations, 285; destroys unfavourable variations, 286; all species produced by, 287; many domestic animals protected from this law, 288, 291; theory requires that there should be a series of intermediate types graduating into one another, 297; does not explain why variations occur, 300; evidence in favour of this law, 301-6.

Odontornithes, 304.
Oldenburg, Mr, 33.
Opossum, fossil, 164, 165.
Organisms, how produced, 272.
Organs, law of correlation of, 159-161.
Origin of species by means of Natural Selection, 223-228, 292, 300; Darwin's book on, 280.
Ornithology, 186.
Orohippus, 306.
Ostracoderma, 47.
Oviparous animals, 47.
Owen, Sir Richard, *note*, 88; publishes descriptive catalogues of the Hunterian Museum, 89; on Cuvier, 150; on the correlation of organs, 161.

Palæolithic men, 2.
Palæontology, 9; Cuvier the chief founder of, 152; relations of this science to geology, 153; Cuvier gave scientific bases to, 158; enables us to trace line of descent of animals, 306.
Peacock, 3; black-shouldered, 299.
Pennant, Thomas, account of, 96; on the migration of the herring, 98, 99; on the

common hog, 100, 101; on the wild boar, 102; tours and publications, 102-106; on migration of swallow, 119.
Petiver, James, the collection of, 68.
Peyssonnel on zoophytes, 94.
Physiology, 8.
Pig, domestic, relieved from law of natural selection, 288, 290.
Pigeon, an extinct, 68.
Pigmentum nigrum, in animals, 79.
Polypes, freshwater, 94; trumpet, 191.
Primates, 58.
Primrose, the, 215.
Protective resemblances among animals, 229-232.
Pterodactyles, 163, 304.

Quadrupeds, division of, by Linnæus, 58; primates, 58; embryo of, 302.
Quinary classification, the, 177.

Radiata, the, of Cuvier, 147.
Raven, 50.
Ray, John, 18; the chief representative of natural sciences in pre-Linnean period, 20; his birth and education, 22; tours, &c., 24-33; *History of Insects*, 35; character, 36; botanical treatises by, 37; theological treatises, 38; zoological treatises, 39; his classification of animals, 47, 48; his classification artificial, 48; separates animals which are closely allied, 49.
Reaumur, 95.
Red Lion Club, the, *note*, 199.
Reindeer, 2.
Reptiles, 47; *History of*, 186.
Retrogression, 168-182.
Richardson, Sir John, 185.
Robinet on man and animals, 156.
Robinson, Dr Tancred, 33.
Rondeletius, 21.
Rook, the, 50.
Royal College of Surgeons, 88.
Ruminant animals, 303, 304.

Sainfoin, the, 215.
Sargassum, 222.
Sars, the Norwegian naturalist, 191.
Scallop, the Icelandic, found in the Clyde, 219.

Scandinavian area, 215.
Scout, the, 26.
Sea-squirts, a link between the true shell-fish and vertebrate animals, 301.
Selection, artificial, as expounded by Darwin, 289.
Serpent worship, 4.
Sexual selection, theory of, 232.
Shaw, George, 108; chief works, 109.
Shells, Lister on, 91.
Sibbald, Sir Robert, 91-93.
Sloane, Sir Hans, 33, 54, 65; his *Catalogue of Jamaica Plants*, 66; founder of the British Museum, 67.
Smeathman, entomologist, 108.
Solan geese, 25.
Solander, 55.
Sowerbys, the, 188.
Sparrman, 55.
Species, Forbes on fixity of, 206; natural groups of, 208; definition of, 244; transmutation of, 236-263; old views as to, now given up by naturalists, 236, 237; revolution accomplished by Charles Darwin, 237; same principle of evolution adopted by Lamarck, 237; earliest definite theory of evolution in Lamarck's *Philosophie Zoologique*, 243; many naturalists believed that species of animals and plants were special creations, 243; general ideas that a species consists of an assemblage of individuals, 244; difficult to decide between species and a variety, 245; the physiological test of, 245; Lamarck on, 246, 247; only stable so long as their environment remains unchanged, 247; agencies concerned in the modification of, 251; giraffe theory, 252; old idea that species produced as we now find it, 254; notion of permanence of, 257; change in structure of organs from use or disuse, 259, 263.
Spence, William, 189.
Spencer, Herbert, and the survival of the fittest, 284.
Stag, use of horns of the, 230.
Star-fishes, 204.
St Paul's battoons, 33.
Stephens, James F., 189.
Swainson, William, 91, 168; travels

in South America, 170; issues his
Zoological Illustrations, 170; theory
of circular classification, 174-179; circular classification a mere figment, 182.
Swan, the, in folk-lore, 3.
Swine, nose of the, 230.
Systema Naturæ, publication of, by
Linnæus, 53; described, 55; *note*, 57.

Tamarisk, the French, 214.
Taxonomy, 9.
Teleology, 10.
Terrestrial animals, 47.
Termites, 108.
Tessier, Abbé, 137.
Testacea, 47; Montagu on, 107.
Thylacinus, the, of Tasmania, 166.
Tongue, the use of, in cattle, 230.
Torpedo, John Hunter on, 78.
Tortoise, described by White, 114.
Totemism, 4.
Tradescant, John, museum of, 68.
Transmutation of Species, 236-263.
Trembley, Abraham, notice of, *note*, 94.
Trout, gizzard or gillarroo, 79.
Trumpet-polype, 191.
Typical group of animals, 180, 181.

Ungulatæ, 59.

*Variation of Animals and Plants under
Domestication*, by Darwin, 280.
Vegetables, definition, by Linnæus, 58.
Vermes, classification, by Linnæus, 61.
Veronica, the alpine, 215.
Vertebrata, the, 47; division of, by
Linnæus, 56; Cuvier on, 147; researches on, 153.
Vestiges of Creation, by R. Chambers, 264.

Viviparous animals, 47.

Wallace, Alfred Russell, joint-author
of the theory of natural selection, 281.
Westwood, Professor J. O., 189.
Whale, the, 79, 173, 303; Sibbald's, 93;
Cuvier on, 150.
White, Rev. Gilbert, 110; account of
his life, 111; *Natural History and
Antiquities of Selborne*, 112; account
of 'Timothy,' a tortoise, 114; on the
flight of birds, 116; on migration of
the swallow, 119.
Willows, the dwarf, 215.
Willughby, Francis, 20; friendship with
Ray, 39; joint tour with Ray, 40;
early death, 41; editing of his scientific
notes by Ray, 42; his ornithology, 43;
his writings, 43-45.
Wilson, Alexander, 'the American ornithologist,' 121; pedlar and poet, 123;
emigrates to America, 124; life as a
schoolmaster, 125; introduced by
Bartram to the study of natural
history, 125; *American Ornithology*,
126-128; characteristics of, as a
naturalist, 128; his description of the
osprey, 128.
Wolf, 3, 79.
Wombat, jaw of, 163.
Woodward, Samuel, conchologist, 188.

Yarrel, William, 187; his natural history
of birds and fishes, 187.

Zones of depth, the four, 210.
Zoologists, British, 90-135.
Zoology of the Voyage of the Beagle, by
Darwin, *note*, 278, 280.
Zoophytes, 94, 189; *History of*, 190.

THE END.

Edinburgh:
Printed by W. & R. Chambers.

www.ingramcontent.com/pod-product-compliance
Lightning Source LLC
Chambersburg PA
CBHW030736230426
43667CB00007B/738